The Car Design Yearbook 6

Stephen Newbury

The Car Design Yearbook 6

the definitive annual guide to all new concept and production cars worldwide

MERRELL
LONDON • NEW YORK

Contents

Trends, Highlights, Predictions

Trends, Highlights, Predictions

With the perspective that comes from our current vantage point, nearing as we are the end of the twenty-first century's first decade, we can see that the constant compulsion of manufacturers to innovate is no less strong this year than it has been in any other recent year. Innovation, by its very nature, never stands still: the focus of the auto industry's inventive efforts switches from one area of technology to the next as the years roll by. And likewise, the top manufacturers lead by turns in different areas of expertise as the balance of capability shifts from one set of engineers to another; there may even be the occasional surprise to disrupt the status quo – such as when Honda unexpectedly became the most skilful of all companies in making its vehicles friendlier to pedestrians.

In the 1970s, which saw oil crises and dramatic fuel price rises, major strides in powertrain technology heralded the modern era of advanced and efficient engines; the following decade, prompted by the rise of the Green movement, especially in Germany, saw an equally concentrated focus on controlling tailpipe emissions. The 1990s, meanwhile, may go down in automobile history as the decade the diesel engine came of age, and when electronic systems first gained an important role in overseeing the dynamics of our vehicles.

We are now going through a second revolution when it comes to powertrain innovation. This time, however, much more of the attention is directed to the development of alternative propulsion systems. Hybrid cars of varying descriptions are now marketed by half a dozen companies, with many more set to launch their offerings in the near future; conventionally engined cars capable of running on biofuels and other green fuels are now commonplace in all manufacturers' model line-ups, although the actual use of these fuels will not become widespread until their availability in filling stations is improved and governments harmonize the fiscal incentives intended to promote their adoption.

In this respect the wave of company consolidations that swept through the industry in the 1990s has proved beneficial. These new, larger industrial units – whether among carmakers or their top-level suppliers – provide the critical mass to enable the development and sharing of new fuel technologies to occur both more swiftly and more economically. With development teams now sharing knowledge, global vehicle platforms can be designed from the very outset to accept hybrid drives. The same applies to component makers: in Germany, for example, the once hybrid-dismissive premium car establishment will shortly be able to hit the ground running with off-the-shelf hybrid packages from such leading suppliers as Bosch, Continental and ZF. This gives the opportunity for technology to be quickly shared among the car manufacturers.

Impelled mainly by government policy and public social conscience rather than the cost of oil, ecologically responsible cars are the big growth area we can see today. Nowhere has this been demonstrated more plainly than in North America, where big gas-guzzling SUVs are sitting

Above
The Citroën C-Métisse concept shows how the French manufacturer is determined to design ever more voluptuous style into its cars.

Right
The Venturi Eclectic throws some practicalities out of the window in return for incorporating eco-friendly technologies to create a low-carbon vehicle.

unsold in dealers' lots while Toyota and Honda struggle to build enough Priuses and Civics to satisfy booming consumer demand. In this edition of the *Car Design Yearbook* we take a much closer look at these so-called eco-cars, explaining the alternative technologies available today and attempting to summarize the reasons why climate change is now focusing the minds of responsible governments and manufacturers.

We also take a detailed look at glazing trends and technologies. More glass is being used in the construction of cars than ever before, and there are some interesting developments in the technologies that can be built into the glass panels: ultraviolet reflectivity, automatic demist, communications antennas and automatic lighting systems can all be designed into what used once to be a humble sheet of glass. Each of these enhances the technical functionality of the car and improves passenger comfort; and our article also explains how, implausible though it might seem, glass can even improve the fuel economy of a vehicle.

It has been our tradition in previous editions of the *Car Design Yearbook* to profile leading designers working in today's auto industry. This edition is no different, and here we highlight the stories of three of the most influential people in the business: Gerry McGovern, who designed the Land Rover Freelander and the MG TF; Audi veteran Peter Schreyer, who is now charged with building a global visual image for fast-growing Korean automaker Kia; and Frank Stephenson, whose credits include the Mini and the Ferrari F430. We document how they came to be leaders in their field, the models that have shaped their careers, and the challenges they are facing today. We even have the low-down on which models motivate them as designers

and which other designers have had the biggest influence on how they operate.

The annual calendar of international motor shows has, as always, produced a healthy harvest of new concept and production cars from the world's automakers. The American shows in particular appeared full of sporty designs, although conspiracy theorists would most likely argue that these very media-friendly cars were nothing but a rearguard action of the heavy-metal brigade in the face of the growing popular appeal of more pro-environmental cars. On the other hand, designers are at the same time using different 2007 concepts to remind us that eco-fuelled vehicles can look exciting and be fun to drive. The Toyota FT-HS, providing supercar performance from a comparatively low-cost hybrid system, and the Chevrolet Volt, with its battery drive topped up by an ethanol-fuelled charging engine, are both examples of cars that combine green with keen.

This year has been another great year for design in general. Since the beginning of the decade car designers have spent much of their time and energy focusing on improving the interior design of cars, an aspect of the vehicle

Opposite
One of Mazda's latest concepts, the Ryuga, is based on the idea of *nagare*, which means 'flow' in Japanese. Even the pearlescent paintwork is meant to bring to mind flowing lava.

Above
Performance-car aficionados often dismiss environmentally responsible cars as boring and slow, but Toyota's FT-HS holds out hope that hybrid technology could one day bring blistering performance.

that had been lagging behind new contemporary exterior shapes. The effect has been remarkable: even a low-cost Skoda or compact Fiat now has a well-executed, high-quality interior. Recently, however, the emphasis has begun to shift back to the exterior again, in many cases to achieve a clearer differentiation of one brand from another. This key new area of brand differentiation can be most easily understood by comparing the rear designs of a range of concept cars. Along with the shape of the frontal grille, the rear-end signature is an area where designers strive to communicate that unique brand identity.

Where innovation has gained, rather than lost, momentum in interior design is in the exploration of new lighting technologies within the cabin. The new Mini features mood-setting background lighting, with which the driver can choose a colour to suit his or her frame of mind. This is achieved through the subtle use of coloured LEDs. There have even been concept cars that automatically adapt the colour ambience to reflect the driver's mood, which is measured by monitoring the way in which the vehicle is being driven.

Another area of innovation definitely set for production is the move to digital touchpads to replace conventional electrical switches. These will first be seen on dashboards, but there is no reason why, eventually, interior door panels should not incorporate touch screens for controlling the door latch, window, mirror and security functions. This technology is very appealing to designers as it allows them to specify a completely smooth surface or even one with the same texture as the surrounding trim. With its deliberately iPod-like centre console design, the Volvo XC60 previews one such example of a touch-panel screen.

Over the six years that the *Car Design Yearbook* has been in existence we have documented the steady rise in the numbers of sport utility vehicles being designed, launched and sold. This year, with the key US market having abruptly turned its back on the most extravagant and fuel-thirsty of these, SUV design

Above
Peugeot's overtly sporty 908 RC concept, with its four chrome-tipped exhausts and clearly highlighted diffuser, struggles to marry racy cues with the packaging required for a four-seater cabin.

Opposite top left
The Acura Advance has a distinctive centre line from which bold surfaces emanate, bounded by crisp lines and sharp features.

Opposite top right
The Skoda Joyster's rear lamp cluster helps to give a real sense of playfulness to the car, making it live up to its name.

Opposite bottom left
With its rear screen running beneath the rear spoiler and lights that look like electrical current flowing through wires, Chevrolet's Volt makes the clear statement that this is a design looking towards the future.

Opposite bottom right
The distinctive shoulders created by the rear lamps on the Volvo XC60 concept give an impression of strength and solidity, characteristic of the values of the brand.

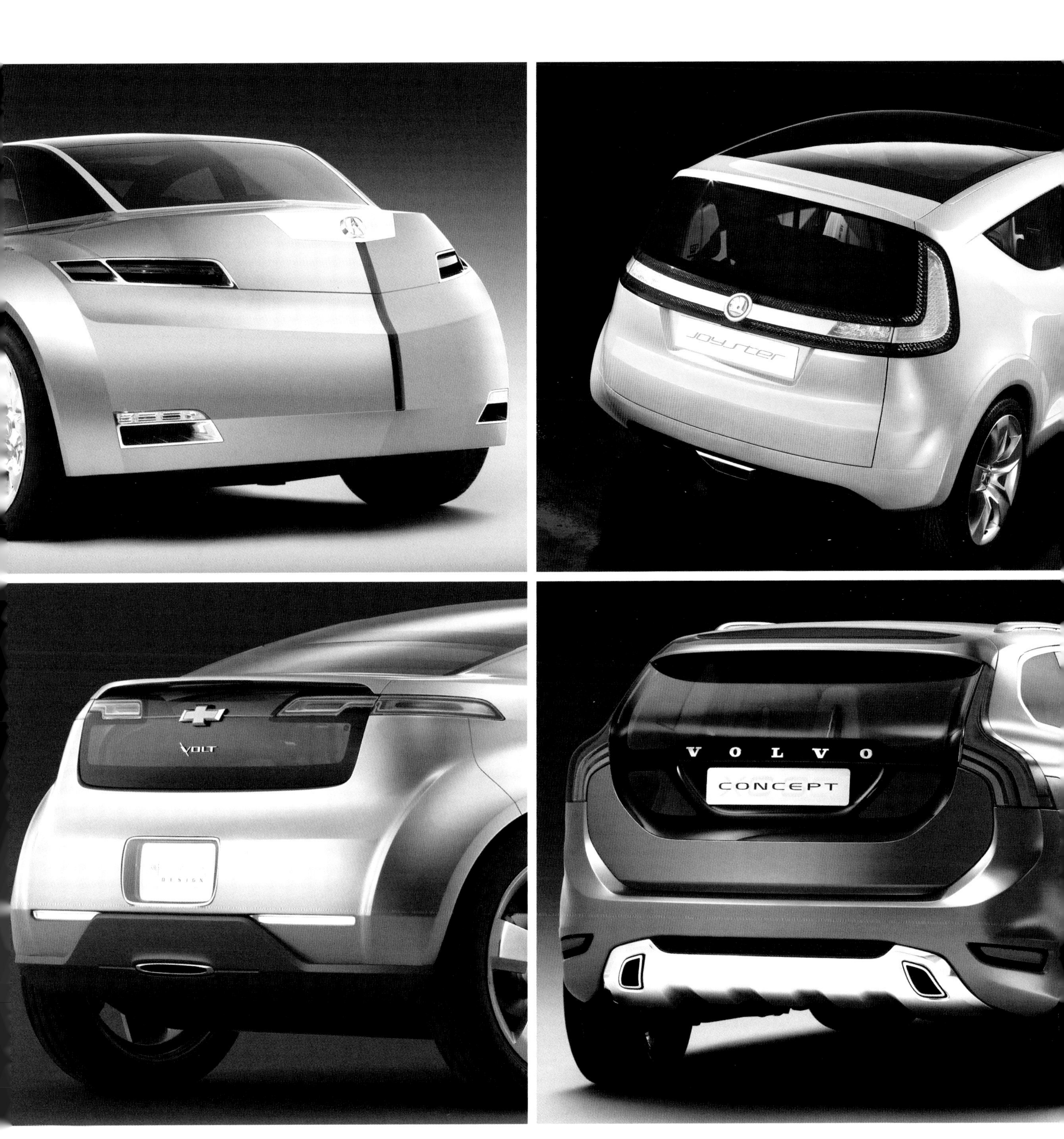
Joyster
VOLT
VOLVO
CONCEPT

finds itself at something of a crossroads. Yes, there is a selection of new SUVs this year, but the messages are mixed. The new Land Rover Freelander, medium-sized by European standards but counting as entry-level in North America, has gone decidedly upmarket; BMW's new X5 is deliberately styled to disguise the increase in its size; while such models as the Saturn Vue, the Buick Enclave and the Acura MDX are so visually constrained by their grand size that it seems almost impossible for the designers to come up with anything original to get excited about.

While such manufacturers as Land Rover are launching more contemporary designs with flat, planar panels and clean, boxy proportions, others are looking to differentiate themselves from the crowd by applying a modern twist on contemporary design and mixing heritage with modern features. Examples are the Alfa Romeo 8C Competizione, the Renault Nepta concept and the Volvo C30: each of these hints back to past decades as well as looking forward.

Peter Horbury, the ex-head of Volvo design who today heads Ford's North American design division, has now got his feet firmly under his Detroit desk. One of the first new concepts to emerge from his studio is the notable Airstream, a modern take on the great Airstream trailers dating back to the 1930s. Using superficial rivets that give a nod to aircraft construction methods, this design is more than just a highly organic form: it packs an environment-friendly hydrogen fuel-cell hybrid powertrain as well. With an interior inspired by 1960s furniture, this concept is designed to try to bring some style credibility back to Ford.

A model that embraces organic forms in a more extreme manner is the Hyundai HCD10 Hellion. This highly polarizing design has extreme animal-like surfaces and bulging exoskeletal riblike structures that mark it out from a wide selection of more conventional designs that sit more comfortably with current trends. This model may provide inspiration to other designers who are searching to position new models much

Above
Clean, fresh and classy design will help the Land Rover Freelander move upmarket – a sound business strategy considering the significant success of the new Discovery.

Opposite top
Renault's Nepta concept car seeks to explore design themes for a long and sleek luxury avant-garde cabriolet.

Opposite bottom
Well-heeled car enthusiasts with a sense of history have rushed to buy the retro-inspired 8C Competizione coupé from Alfa Romeo.

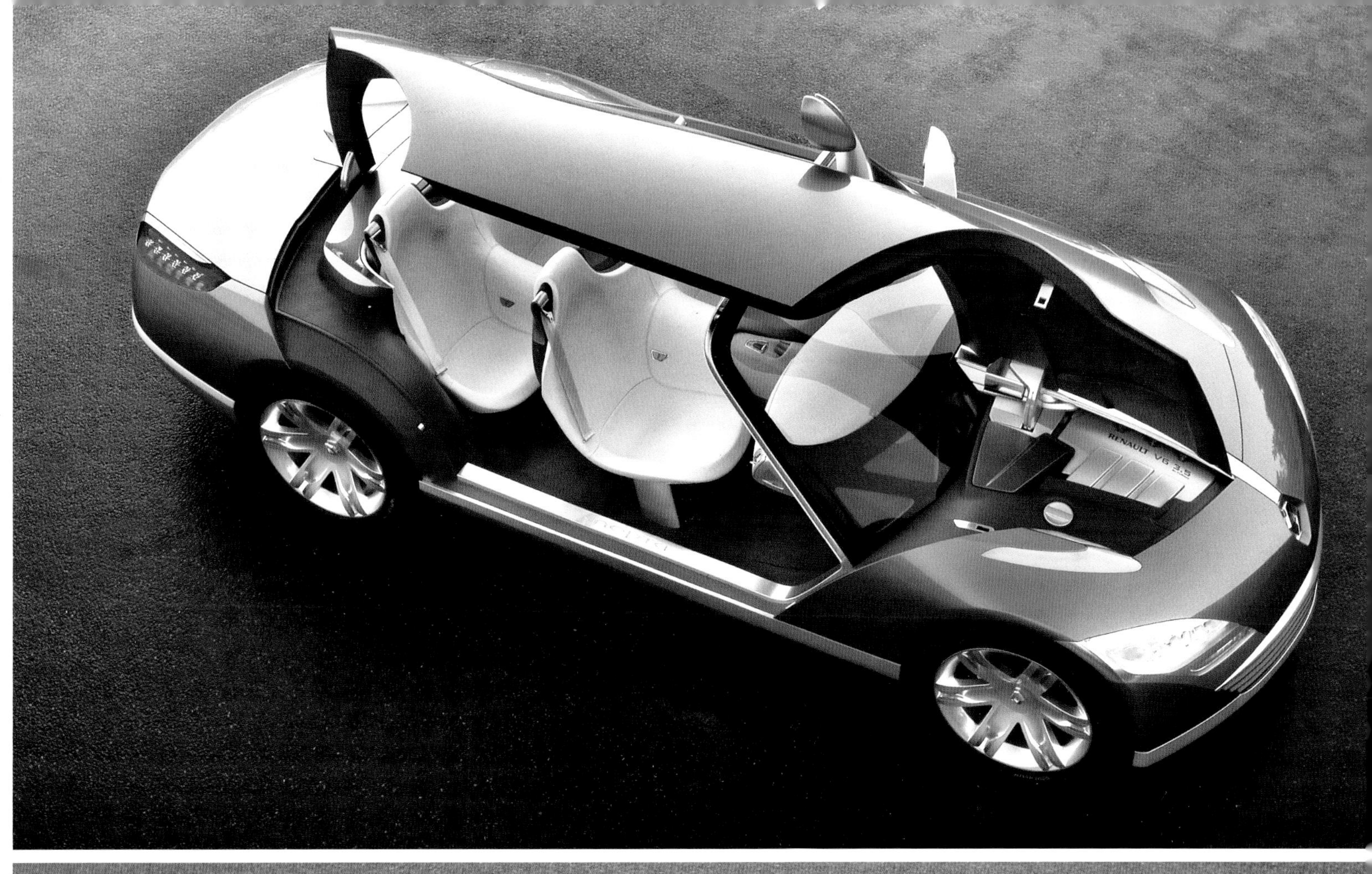
RENAULT V6 3.5

closer to nature, but many customers would feel very self-conscious in such a vehicle.

With the world auto industry facing unprecedented structural change as China's vehicle output grows and India tools up for serious-volume car production, many industry analysts expect the pressure on the major Western companies to intensify rather than diminish. A major shakeout of surplus capacity, outdated factories and unwanted models is already beginning to gain momentum in North America, the region where mass production first took hold. Polarization is beginning to become evident, not just between continents and nations but also between individual carmakers and even individual model lines. From the point of view of design policy, this leads to some interesting observations: while manufacturers with successful models are being extra-cautious in their approach so as not to endanger the strong market appeal they already enjoy, some weaker producers are resorting to more extreme designs in order to attract attention and distinguish themselves from the competition. Examples are the new Mini and the BMW X5, both of which are barely distinguishable from their outgoing shapes, and the powerful, progressive design of the Jaguar C-XF concept and the strong lines of Hyundai's i30 hatchback for Europe.

The booming Chinese auto economy poses something of a conundrum for global carmakers. As incoming Western manufacturers have found out to their cost, Chinese consumers tend to prefer cars that are designed specially for their market, to reflect peculiarly Chinese tastes and requirements. This conflicts directly with the international automakers' firm preference for unified global car designs, designed and engineered centrally in order to maximize returns on their development costs, but built globally in satellite plants nearer to the consumer and with only minor changes to colour and equipment as a concession to local tastes. Luxury cars in China are mostly chauffeur-driven, meaning that the focus shifts from comfort in the front to space in the rear. At the other end of the scale, tiny multi-purpose vans are big business in China, too: these versatile jacks-of-all-trades may be asked to transport farm produce during the week and passengers over the weekend, and perhaps even take livestock to market on a regular basis. There

is nothing in a Western automaker's inventory that can get anywhere near such requirements. The Beijing and Shanghai auto shows will become increasingly important events in years to come; in 2006 Volkswagen, Cadillac and Nissan saw fit to launch wholly new models there.

This polarization is set to increase as huge numbers of individuals – especially in Asia and South America – mobilize themselves through the purchase of a private car. Renault is already enjoying big success with its Logan 'back-to-basics' sedan, with the range expanding to include a station-wagon and a pickup; now other carmakers are looking for ways to bring low-cost vehicles to emerging markets, and India's Tata is well advanced in the development of a still lower cost vehicle that seeks to bridge the gap between the motorcycle and the cheapest cars – and thus mobilize yet another million-strong wave of first-time car owners.

This contrasts plainly – and perhaps painfully – with the super-affluent European, Japanese or American customer shopping for a top-of-the-line luxury vehicle. The choices there may be between walnut and maple for the interior trim, or between carbon and steel for the brakes, or whether or not to tick the option box for radar-guided cruise control that can automatically bring the car to a halt when it detects an obstacle in its path. And it is the infinite variety of such choices and the ever-expanding spectrum of vehicle formats that continue to make car design one of the most inventive and most inspirational product-creation processes in the world today.

Opposite
Ford's association with Airstream, makers of America's iconic streamlined aluminium caravans, is a bid to add heritage value to an ultra-modern concept: here, Airstream's trademark rounded style is used to dramatic effect.

Above
The widely acclaimed design for the Jaguar four-door sports coupé concept, the C-XF, has allowed the marque's designers to look to the future without being tied to past templates.

A Touch of Glass
Green Machines
AIRSTREAM
Ford

A Touch of Glass

In car design, as in any other area of design, fashions come and go. But one of the clearest and most consistent trends that can be identified over the course of the past few decades is the much greater area of glass that is now being introduced into today's vehicles.

Automakers have always tried to maximize the size of the windows on their designs, but engineering limitations prevented them from going as far as they would perhaps have liked. Recently, however, computer-aided design (CAD), engineering optimization and improvements in steel have allowed pillars to become thinner but at the same time stronger, opening the way to a higher percentage of glass in the car's superstructure. Most importantly of all, major advances in the technology of the glass itself have enabled glazed surfaces to be curved as never before, to be used in areas where glass has never been seen, and to perform a whole spectrum of secondary tasks – communication and security are examples – that benefit the vehicle as a whole.

Designers are keen on maximizing glazed area for a variety of reasons. Most obviously, large windscreens and side and rear windows contribute to improved visibility and thus better safety; secondly, the body designers can use interesting glass shapes as important graphic elements in the car's exterior style. The other big change promoting the use of glass has been the shift in overall vehicle shape, as mass-produced family cars have moved from conventional three-box profiles to larger one-box people-carrier-type designs. These generally have very much larger windscreens, often set at a very fast rake, and have lower waistlines and deeper side windows as a result.

More recently still, and perhaps as an extension of the MPV phenomenon, a further key trend has emerged: all-glass roofs, where the bulk of the area that was once sheet metal now consists of a full-length, full-width bonded-in glass panel. These complete glass roofs are entirely different from the small tilt-and-slide panels fitted to many cars in the 1980s and 1990s. All-glass roofs are especially appreciated by rear-seat occupants, who can now benefit from the same light levels as those up front. Increasing the light in a space is a well-known means of improving people's mood and making them more alert: just look at the way in which modern architects rush to build such public structures as airports and hotels with large glass windows and big airy spaces. Car designers and engineers have now woken up to the same thing. The trend can perhaps be traced back to the 1999 Smart City Coupé, which offered the option of a transparent roof panel in lieu of the standard opaque plastic component. Now, from Mini to Mercedes, the glass roof is a staple of the options list.

But while these innovations in glass technology have given designers new freedoms in terms of interior luminosity and ambience, such specialist glass manufacturers as Pilkington, Saint-Gobain Sekurit, Asahi Glass and Nippon Sheet Glass have had to work still faster to develop innovative technologies to deal with

some of the penalties associated with large glazed areas. These include increased weight, greater noise transmission from the outside, and the build-up of interior heat as a result of solar radiation.

Solar gain is a big problem in any glassy structure, but in the comparatively intimate confines of a car cabin it can be a major headache. Air-conditioning systems can have an uphill struggle absorbing the heat that comes through from the sun, adversely affecting fuel consumption (and thus CO_2 emissions) as well as passenger comfort. The latest electro-chromatic glazing systems can automatically adjust the amount of light (and thus solar heat) coming through the roof panels to ensure thermal comfort and minimize the load on the air-conditioning system.

The last decade has seen significant improvements in the design and application of solar-control glazings in vehicles. Solar radiation is partly reflected, partly transmitted and partly absorbed by glazing, the degree of each depending on the specification of the glazing fitted. Coatings for windscreens are now possible that reflect more than 30 per cent of the sun's energy (more than five times the rate of a standard glass). This benefits the latest

Opposite
On the Saab Aero-X concept, the wraparound windscreen and side glass are incorporated together in a lift-up canopy.

Above
Maybach uses an electro-transparent panoramic roof, with passengers given options for adjusting the amount of light that enters the interior. The roof normally provides a view of the sky, but can be made opaque by pushing a button that sends an electrical current to a layer of liquid-crystal-filled film made from electrically conductive polymer material.

generation of vehicles designed with larger glass areas. The coating technique can be used on complex-shaped windscreens and therefore provide the required optical quality. Solar-reflective glass can use an infrared-reflecting film within the laminate or coatings applied directly to the glass surface itself, and reflective glazings have now been successfully integrated into large-area rooflights.

Conductive surface coatings, in addition, make it possible for designers to integrate ambient lighting into the glass, while in the future it should be possible to create localized light sources or full area patterns.

In terms of integrating communications technology, glass plays a major role. Car manufacturers have already moved away from conventional rod-based antennas to improve styling aesthetics and defeat vandals. Glazing companies have helped by developing methods for fully integrating antenna systems either on the glass surface or inside the glazing construction. Head-up display (HUD) systems have long been used in military aircraft to project information into the pilot's field of vision. These systems are now in place in selected cars – for example, certain BMWs and Citroëns – at a cost acceptable to the market.

Windscreen technology has seen most innovation over the years. Defrosting and demisting windscreens have been commonly used for around twenty years. The technology works by incorporating fine wires into the glass ply that are capable of de-icing a frozen screen. This system also adds demist capability to keep the inside of the glass clear, too. Sensors fitted to

windscreens can detect external moisture levels and automatically activate the windscreen wipers. This rain sensor, attached to the interior of the windscreen, operates through a system of infrared light-emitting diodes; it is now commonplace on family cars, especially those from France. Another technology that improves driver visibility is a hydrophobic coating, which is applied to the outside glass surface and significantly improves water-droplet flow from the vision area. Hydrophobic products are used extensively in Japan, with market interest in Europe and in North America: Audi used this technique on the very raked rear screen of its A2, thus avoiding the need for a rear wiper.

Moving to rear screens in general, a number of new models are featuring what appear to be all-glass tailgates. In fact, metal structures are often bonded on the inside of the glass, hidden by the black obscuration band. Examples include the new Volvo C30, the Citroën C1, the Peugeot 107 and the Toyota Aygo.

Glass panels are increasingly being specified by the growing band of companies specializing in

Above
In the Renault Clio Grand Tour concept, the windscreen and roof are made from a single glass panel, and the front header rail has been removed, thereby creating for the occupants an amazing sense of being outside.

Above
The Renault Zoé concept has starlike lights built into its glass roof to illuminate the cabin at night.

Opposite
The Peugeot Moovie concept shows just how far designers are prepared to go when exploring the use of glass.

folding convertible roofs for coupé-cabriolet derivatives. The complex folding mechanisms already allow for interesting configurations of sliding and folded openings, but the Volkswagen Eos goes one better: its folding hard-top section incorporates a glass panel as part of this roof system, so that even when the roof is up, light and air can still flood in if the sunroof is tilted or retracted. The Renault Mégane coupé cabriolet, which has been on the market for some years, was the first to have a folding roof incorporating a glass upper panel.

The trend towards the use of more and more glass clearly conflicts with the desire for lighter vehicles. Carmakers have tasked glass manufacturers with developing lighter-weight glass panels that display the same acoustic and security characteristics as thicker panels and that are just as strong. Bonded-in windscreens and, increasingly, rear windows contribute a surprisingly high proportion of a vehicle's overall torsional rigidity; a reduction in glass thickness can thus have a direct bearing on a vehicle's ride performance.

Acoustic isolation was once famously tackled by Mercedes-Benz with clumsy double-glazed side windows in its 1990s S-Class limousine. Techniques have improved enormously since then, to the extent that Jaguar, in its 2005 XJ 2.7 diesel, was able to isolate the high-frequency clatter of the diesel engine by the use of special five-layer 5-mm (0.2-in.) glass from Sekurit in all the side windows. This gives a clearly discernible seven-decibel reduction over the already good standard laminated glass.

Since 1970, the standard glazing constructions for cars have been laminated glass for the windshield and toughened glass for the side and rear windows. Laminated glass is not only stronger and safer (it is less likely to shatter) but also much more resistant to attack from vandals and thieves. For these reasons it is expected that over the next ten years side windows will move from toughened glass to laminated glass. This represents a turning point for advanced vehicle design opportunities: now designers may exploit the ability of the glass to become opaque or non-transparent to solar radiation; they can add interior light effects, wire in antenna systems and even potentially add exterior signalling or messaging functions.

Laminated side windows have a similar construction to laminated windscreens, namely a PVB interlayer sandwiched between two glass plies. In order to meet door-slam tests, both plies are semi-toughened to give additional strength to the glazing. The trend for lighter-weight glazing means that an overall 4-mm (0.16-in.) makeup is becoming critical as laminated systems are adopted on smaller, lower-segment vehicles. Acoustic performance can also be enhanced by the use of special interlayer materials.

The dimensional tolerance of glass is vital to the overall quality of the vehicle. Badly fitting glass can cause water leakage, wind noise, poorly fitted trim or difficulty in opening windows. Glass providers have developed computer simulation programs for the glass-forming process, allowing them to work with designers to create new shapes that will be feasible for eventual manufacture. For example, optical properties of curved windscreens can be virtually assessed at the concept-vehicle stage without expensive tooling being made.

The trend towards suppliers to the automotive industry gradually increasing their remit has not escaped the glass industry. Glass suppliers are now providing complete systems that extend beyond the glass to include the fixings and trim panels around glass. This allows a very accurately controlled assembly that can include watertight seals to give better water management and easier and more accurate assembly to the vehicle.

All this, of course, is no more than would be expected by today's acutely quality-conscious car buyer. Car designers always strive to go one step further, and now the newly design-aware glass industry is poised to help carmakers build those inspirational new models – such as the Saab Aero-X, with its huge aircraft-like screen canopy – that once used to be dismissed as hopelessly over-ambitious.

Green Machines

For all their appeal as machines of style, speed and sophistication, cars have been bad news in 2006. Not through any fault of their own, maybe, but because, by virtue of their sheer numbers, they have rarely been off the front pages in the growing debate over CO_2 emissions.

Like it or not, cars have been in the firing line of every discussion sparked off by two epoch-defining reports: the first, the Stern report, co-ordinated by a highly regarded senior UK economist; and the second compiled by the United Nations Intergovernmental Panel on Climate Change (IPCC). Both reports focus on carbon-dioxide emissions, global warming and the inevitability of irreversible climate change; both spell out in explicit terms the dangers of the path that industrialized nations are currently on, and both describe equally explicitly the urgency of taking immediate and decisive action.

Although both Stern and the IPCC are scrupulously balanced in their examination of the many different sources of greenhouse-gas emissions, and although their recommendations are also impartial and unbiased towards any particular interest groups, the public debate that has grown up in the wake of the reports has been less objective, focusing on – and demonizing – air travel and road transport. Cars are, in fact, perhaps the most publicly visible manifestation of conspicuous consumption (of fuel and road space) and CO_2 emissions. Road transport as a whole is responsible for around one-fifth to one-quarter of global CO_2 emissions, though private cars account for only a part of this.

Most recently of all, the European Union's proposal to introduce a target CO_2 emissions limit of 130 grams per kilometre for all new cars from 2012 has brought renewed focus on fuel consumption and thrown the automotive industry into some chaos. With a few notable exceptions, such as PSA–Peugeot Citroën, Renault and Fiat, most carmakers have been faced with the realization that they have nothing in the future-projects cupboard that could bring them anywhere near the required average values by the stipulated deadline.

However, looking at the wider picture, the search for a more sustainable solution to the issue of automobile emissions is being conducted on four main fronts: influencing driver behaviour, such as when and where cars are allowed to drive; incentives or penalties to encourage the purchase and use of lower-emission models; the development of greener power sources, such as biofuels; and engineering developments within the car itself. Of these, it is probably only the final aspect that is likely to have a lasting effect on the way future vehicles will look, feel and drive.

Yet already London, with its new congestion charging zone, has witnessed the arrival of a new generation of tiny electric cars, such as the G-Wiz and the Mega City, designed to exploit the fact that zero-emission and ultra-low-emission vehicles are exempted from the daily charge. These toylike two-seaters certainly challenge our established preconceptions as to how a car should look and operate, so here is a clear example of legislation directly influencing

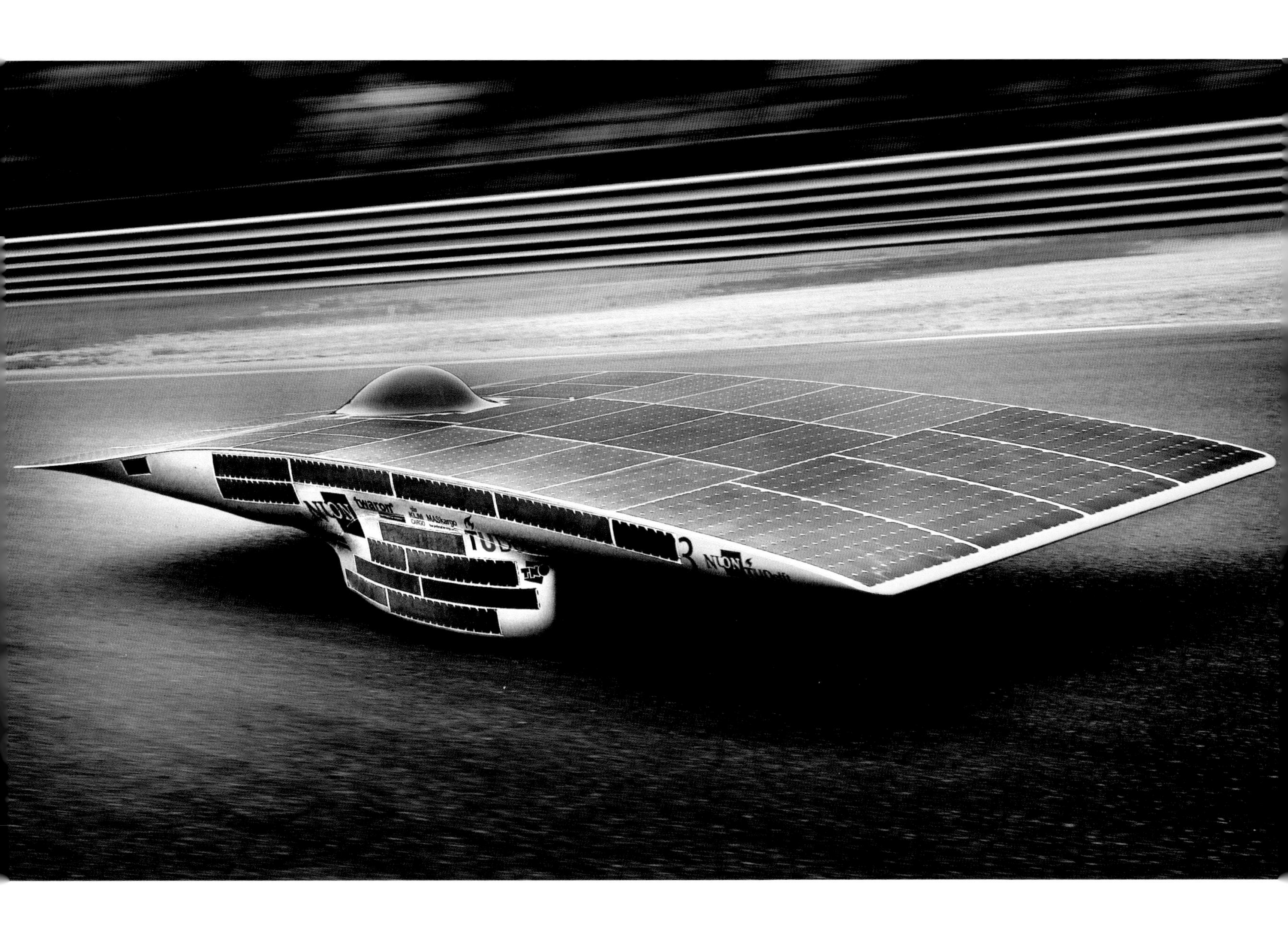

Opposite
Daihatsu's Materia employs an 'intelligent' catalyst, using nanotechnology principles, to improve the catalyst life and thus reduce long-term emissions.

Above
Such seemingly radical vehicles as the Nuna 3, developed by a team of students at Delft University of Technology in The Netherlands, could help pave the way for commercial solar-powered cars. In 2005 the Nuna 3 won the World Solar Challenge race, held in Australia every two years; it achieved a top speed of 140 km/h (just under 86 mph), and covered the 3010 km (1860 miles) at an average speed of 103 km/h (64 mph).

car design. Even Smart has started developing an electric version of its iconic Fortwo for a field trial in London.

In France and Italy, many decades of tax regimes penalizing large and powerful models have led to national car fleets heavily skewed towards small models and diesels – again an example of the way in which incentives and penalties can influence the kind of cars that people buy.

The tailpipe CO_2 emissions of a conventional car are directly proportional to the amount of petrol or diesel – or other combustible fuel – it uses. So the struggle to reduce CO_2 is in effect a campaign for more economical vehicles, but with the added twist of new, greener choices of fuel to give the planners an extra lever to manipulate. Biofuels help trim net CO_2 emissions by virtue of the fact that the crops they are derived from have absorbed CO_2 from the atmosphere while they were growing. (It is worth noting, however, that biofuels may turn out to be a double-edged sword. Some methods of producing biofuels, palm oil in particular, are in themselves highly polluting and negate the carbon-neutral effects of using a non-petroleum fuel; and many biofuel plantations are at the expense of the world's already-endangered rainforests.) Already, up to 5 per cent bioethanol and biodiesel are blended into European petrol and diesel respectively, and there is a European target for 25 per cent by 2030. Sweden's national target is much more ambitious: it wants to eliminate non-bio fuels by 2020, and has already got off to a flying start by giving massive tax breaks and free city parking to cars running on an 85 per cent bio mix. Saab, in particular, has benefited from this and is the only carmaker to offer this type of flex-fuel option across its whole range.

Yet the cars in Sweden still look the same as standard models: it is only what is under the bonnet and in the tank that has changed. However, two propulsion options that promise to bring dramatic changes to the technical layouts of cars – and thus their external proportions and appearance – are the related technologies of

Above
The NICE Mega City runs on battery power and is targeted at London drivers. It was developed from a petrol model made by the French quadricycle maker Aixam.

Opposite
In London, which is seen as the city most open to electrically powered vehicles, a fleet of Smart Fortwos using an electric powertrain (below) is undergoing a large-scale field trial.

batteries and fuel cells. By powering the wheels from motors mounted directly in the hubs, the electrically driven car can do away with the mechanical drive shafts, axles, differentials, gearboxes and clutches that constrain the layout of a standard diesel or petrol car; the batteries – or in the case of the fuel-cell car, the cell stack that provides the electric current – can be placed anywhere that is convenient to the designer. Even the hydrogen storage for the fuel cell can be conveniently secreted within the car's structure.

The opportunities for designers are enormous: no longer will the shape of the vehicle have to be drawn around the template of fixed 'hard points' dictated by the standard positioning of engine, transmission and driveline; interiors, likewise, could be transformed once electrical 'by wire' systems replace such familiar mechanical control interfaces as the steering wheel and foot pedals. Already, moves have been made in this direction, with the introduction on some models of electric parking brakes – requiring only a small switch, rather than a clumsy lever – and electrically actuated gear shifters, as on the Citroën C4 Picasso people-carrier.

For a foretaste of just how these breakthrough technologies can be exploited to remarkable practical effect, look no further than GM's 2002 Hy-Wire concept, which married the company's earlier skateboard-like Autonomy fuel-cell chassis platform to a roomy passenger superstructure with a flat floor running uninterrupted from front bumper to tailgate. Even GM, however, would concede that such a vehicle is unlikely to appear until 2020 at the earliest: fuel-cell technologies still rate little better than crossed-fingers science-fiction hopefulness in corporate product-planning priorities.

Other renewables are being suggested too, though perhaps not quite so seriously: France's always-original Venturi company, usually best known for its sports cars, showed its Eclectic concept at the 2006 Paris motor show. Shaped like an outsized golf cart, this electric buggy has an array of solar cells on its roof that can recharge its batteries when parked and add 7 kilometres (4.3 miles) to its driving range; it also has a small wind turbine that can be attached when the machine is parked, to top up the batteries further.

Rather higher than wind power on every company to-do list is to grab a slice of the much-hyped market for hybrid cars. Once dismissed as no more than an interim technology on the way to the ideal of hydrogen power, hybrids have been the main thrust of corporate fuel-saving research budgets since the turn of the millennium: soon, a company that does not have a hybrid – or three – within its line-up will be excluded from the environmental equation and receive a commercially damaging hammering in the sales charts.

Ever since the Toyota Prius and the Honda Insight pioneered hybrid driving in the late 1990s, two separate strands of hybrid development have been evident: models that are hybridized versions of standard designs, such as the Honda Civic, and standalone hybrids – exemplified by the Prius – which are dedicated designs

conceived around the hybrid powertrain to maximize its fuel-saving advantages.

While there is no inherent reason why a hybrid car should look any different from a conventional model, Toyota did elect to give the Prius a style that was more advanced and more progressive than its routine ranges. Accepting that this arched look has come to be associated with the hybrid way of doing things, Toyota's chief designer, Wahei Hirai, is now beginning to build a deliberate visual identity for hybrid models to make them more distinctive in the showroom and on the streets; the attractive Hybrid X concept, shown at the 2007 Geneva motor show, is the first move in this plan.

Honda, too, exhibited a concept model, the unimaginatively named Small Hybrid Sports coupé, to attempt to put a shape to hybrid underpinnings. Yet with only four makes – the other two are Lexus and Ford USA – in the hybrid game at the time of writing, it is premature to look for visual trends in hybrid design. Right now, with the notable exception of the Prius, a hybrid shows no more differentiation than a petrol or diesel derivative; by the middle of 2008, however, trend-spotting may be rather easier. By then, possibly triple the number of hybrids will come to market as we witness the fruition of the numerous hybrid programmes that kicked off in 2003 and 2004 as companies realized that hybrids were here to stay.

A rather greater chance of achieving a decisive shift in the shapes of the cars we buy could be triggered by two further technical developments that have recently had motor-show exposure: the plug-in hybrid and the series hybrid. Chevrolet's aptly titled Volt, displayed in Detroit in January 2007, can do almost all of its driving on a high-capacity battery, which powers a motor to drive the wheels. But to give it a generous range, it also has a small combustion engine linked to a generator, which recharges the battery: at no stage does the engine drive the wheels directly. This gives the designers much of the freedom of a pure electric car, with the potential to use wheel motors and to place

Opposite
The Saab Biopower 100 is the first production-based turbo engine designed to run on pure bioethanol, also known as E100.

Above
Toyota is continuing to explore hybrid concepts: the aim of this latest Hybrid X proposal is to explore possible design identities for hybrid-powered vehicles.

the components wherever convenient for packaging or external style.

Ford's Airstream, though cast in the shape of a strange leisure vehicle, also shows the potential of electric drive, although this time with the battery being recharged by a small fuel-cell stack. Again, this is the type of drivetrain that would give designers scope for originality in every aspect of the vehicle's layout and look. But it will be a brave company indeed that goes straight in and takes immediate advantage of all of these unaccustomed conceptual freedoms: more likely is a gradual evolution of shapes – much like the aerodynamic revolution in the 1980s and 1990s – that will draw consumer taste along with it and confront popular taboos (such as the need to cling on to the conventional steering wheel) one by one rather than in a single onslaught.

Toyota's FT-HS concept for a compact but high-powered hybrid coupé gives another indication of the way things could conceivably go. Its overall proportion is clearly and reassuringly familiar as that of a sports coupé, but almost every detail departs from accustomed solutions, and the engineering under the skin is of a type new to the market. Were it to be put into production – and the likelihood is that it will – it would be immediately snapped up by the early adopters as that miraculous combination: an eco-friendly super-sports coupé with the power to perform as well as the intelligence to save the planet at the same time. With the toylike G-Wiz at the opposite end of the price and performance scale and standing out from the crowd for very different reasons, who can say that eco-cars all have to be the same?

Above
Honda's Small Hybrid Sports concept is another design study seeking to establish an advanced look for hybrid models.

Opposite
Ford's futuristic Airstream crossover concept is another hybrid variation, with a hydrogen fuel cell providing power for the electric motors.

AIRSTREAM
Ford

Acura Advance
Acura Advanced Sports Car
Acura MDX
Alfa Romeo 8C Competizione
Audi A5
Audi R8
Audi TT
Bentley Brooklands
Bertone Barchetta
BMW X5
Cadillac CTS
Chevrolet Malibu
Chevrolet Volt
Chrysler Nassau
Chrysler Sebring
Chrysler Town and Country
Citroën C4 Picasso
Citroën C-Crosser
Citroën C-Métisse
Daihatsu D-Compact X-Over
Daihatsu Materia
Dodge Avenger
Dodge Demon
EDAG LUV
Fiat Bravo
Fioravanti Thalia
Ford Airstream
Ford Focus
Ford Interceptor
Ford Iosis X
Ford Mondeo
Ford Mustang
Honda Accord Coupé
Honda Remix
Honda Small Hybrid Sports
Honda Step Bus
Hyundai HCD10 Hellion
Hyundai i30
Hyundai Quarmaq
Hyundai Veracruz
Infiniti G35
Italdesign Vadho
Jaguar C-XF
Jeep Patriot
Jeep Trailhawk
Kia Carens
Kia Cee'd
Kia Kue
KTM X-Bow
Lada C-Class
Lancia Delta HPE
Land Rover Freelander 2
Lincoln MKR
Lotus 2-Eleven
Maserati GranTurismo
Mazda2
Mazda CX-9
Mazda Hakaze
Mazda Nagare
Mazda Ryuga
Mazda Tribute
Mercedes-Benz C-Class
Mercedes-Benz Ocean Drive
Mini
Mitsubishi Lancer
Nissan Altima
Nissan Bevel
Nissan Livina Geniss
Nissan Qashqai
Nissan Rogue
Nissan X-Trail
Opel GTC
Opel/Vauxhall Antara
Opel/Vauxhall Corsa
Peugeot 4007
Peugeot 908 RC
Pontiac G8
Renault Koleos
Renault Nepta
Renault Twingo
Rinspeed eXasis
Saturn Aura
Saturn Outlook
Saturn Vue
Scion Fuse
Scion xB
Scion xD
Skoda Fabia
Skoda Joyster
Smart Fortwo

A – Z of New Models

Acura Advance

To say that the Acura Advance generates a degree of surprise would be an understatement: it is very unusual indeed, and creates a distinct shock when first seen. Yet the same may be true for any vehicle that, like this study from Acura (Honda's premium brand for North America), is intended to anticipate consumer tastes many years ahead.

To today's eyes, the concept has a scarily aggressive look to it: the high waistline, low ground clearance, huge wheels and jawlike grille at the front all combine to generate a rather sinister feel. The black mesh grille even has a cutout at the base, making it appear as if it is a drain for whatever evil fluids might emerge.

The design comprises a series of jarring surfaces that convey a sense of angry geometry. The short occupant canopy, with its blackened pillars, is inset, allowing a sharp-edged shelf to run along the DLO from front to rear to emphasize the values of strength and solidity. The rear wheel arch is more pronounced than the front and is highlighted by a circular crease line. Behind the front wheel arch the curved sculpted area serves to reduce the slablike effect of the deep body sides, and culminates in an air vent to feed the rear brakes.

Luxury accents include such features as the aluminium rims around the lights, and the fitted exhaust tailpipes. Acura quotes its designer, Dave Marek, as saying that the goal of this study was to create 'a sophisticated, refined sedan with a mysterious presence'. We're not convinced they have succeeded in the sophisticated and refined look, but there is no question about the sense of mystery.

This is certainly an eye-catching concept, though perhaps not for entirely the right reasons. Quite how it will inspire future Acura and Honda products remains uncertain.

Design	Dave Marek
Brakes front/rear	Discs/discs
Front tyres	22 in.
Rear tyres	23 in.
Width	2019 mm (79.5 in.)

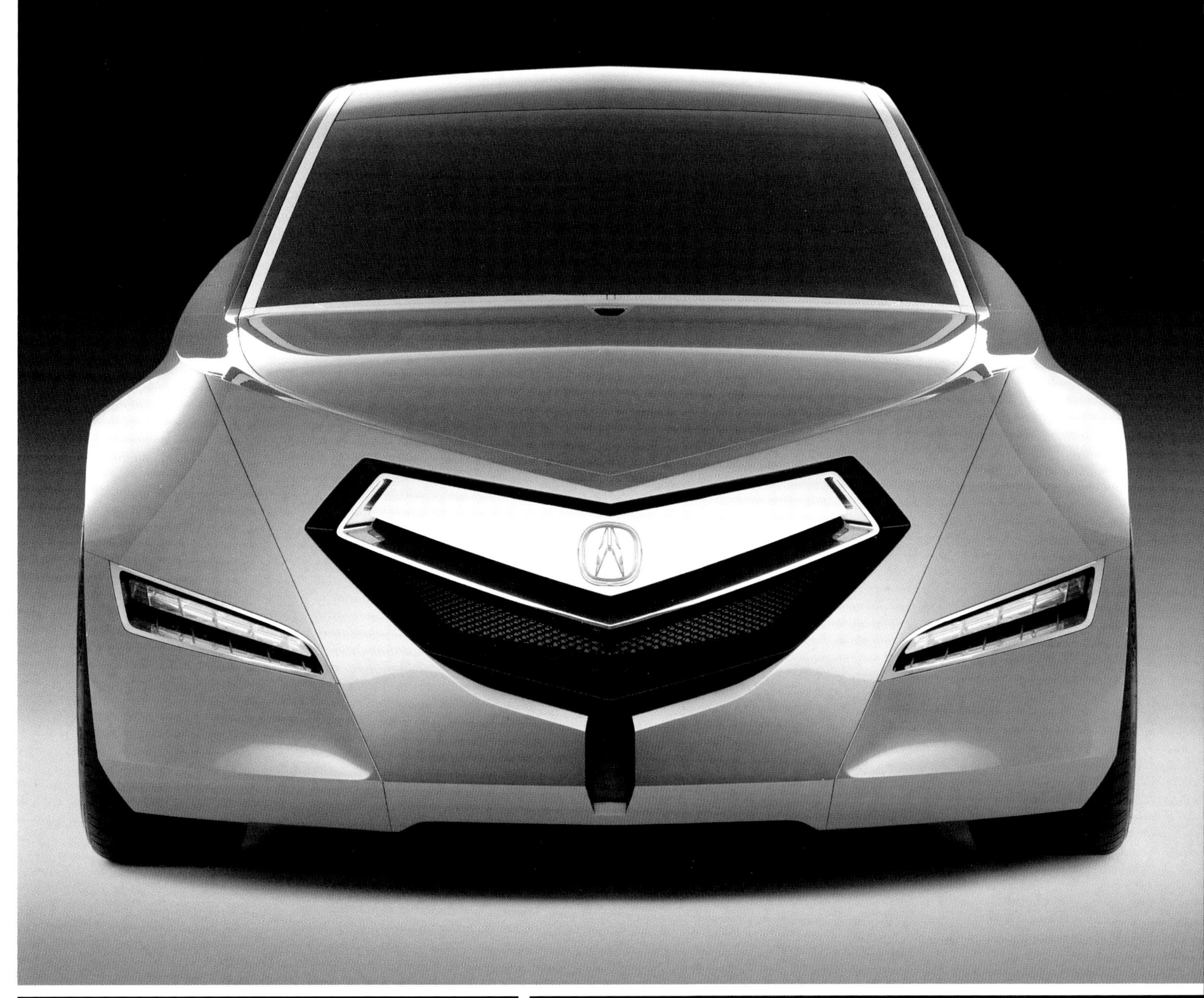

Acura Advanced Sports Car

Design	Jon Ikeda
Engine	V10
Installation	Front-engined/four-wheel drive
Brakes front/rear	Discs/discs
Front tyres	255/40R19
Rear tyres	295/35R20
Length	4610 mm (181.5 in.)
Width	1996 mm (78.6 in.)
Height	1222 mm (48.1 in.)
Wheelbase	2764 mm (108.8 in.)

Acura, the US-only luxury arm of Japanese technology powerhouse Honda, has been facing something of a dilemma since July 2005, when Honda announced that it would stop building the NSX supercar, branded as an Acura in North America. Without the much-admired mid-engined two-seater there has been no obvious performance flagship in the line-up and the role of technology leader has had to be played by the RL sedan.

The Advanced Sports Car concept points towards the production model that will eventually replace the NSX, so it marks a high emotional investment for both Honda and its worldwide fans.

The most obvious contrast with the mid-engined NSX is that, under the ASC show car's lithe and edgy skin, it conceals a front-engined, four-wheel-drive layout. This completely changes the proportions, to the extent that the two models are linked by details, such as the full-width tail lights, rather than their overall impressions.

Although the ASC was shown in Detroit only as a hard exterior model, its design is radically different from what commentators had expected. In some respects it resembles a stealth fighter, the body styling mixing plain surfaces with stiff creases. It is not classically beautiful like an Aston Martin: rather, it communicates tension through the rising body creases that run up over the bonnet and through the doors. There is no central front grille; instead the twin air intakes sit below the fierce-looking LED headlamps. Tight body-to-wheel relationships and a close proximity to the ground make it look securely fastened to the road.

The rear is dominated by complex full-width lamps, well integrated with the rest of the shape, but the rear-window treatment is unconvincing and impractical. Overall, this is a technical design, in contrast to most other supercars, which communicate via a language of organic forms. As such it may not be the car that passionate NSX fans have been waiting for.

Acura MDX

Design	Ricky Hsu
Engine	3.7 V6
Power	226 kW (300 bhp) @ 6000 rpm
Torque	373 Nm (275 lb. ft.) @ 5000 rpm
Gearbox	5-speed automatic
Installation	Front-engined/all-wheel drive
Front suspension	MacPherson strut
Rear suspension	Multi-link
Brakes front/rear	Discs/discs
Front tyres	255/55R18
Rear tyres	255/55R18
Length	4844 mm (190.7 in.)
Width	1994 mm (78.5 in.)
Height	1679 mm (66.1 in.)
Wheelbase	2750 mm (108.6 in.)
Track front/rear	1720/1715 mm (67.7/67.5 in.)
Kerb weight	2070 kg (4563 lb.)
Fuel consumption	11.3 l/100 km (25 mpg)

Luxury SUVs have been an important factor in the North American market for the best part of a decade. Yet Honda, through its luxury nameplate Acura, has always tried to offer a subtly different take on this familiar vehicle format; the first-generation MDX, though now looking dull and dated, projected a lighter, fresher and sportier air than most of the imposingly brutal domestic US products.

Now, with the switch to its second generation, the MDX becomes much more up to date, and indeed makes the original edition look ordinary by comparison. It is slightly wider than the outgoing model, and has a longer wheelbase, while the various parts that make up the new MDX look more integrated. At the front the grille and the complex high-intensity discharge (HID) headlamps run into each other to form an attractive and unusual face, the colouring of the grille matched to that of the lower bumper. From the side, the wheel arches are more prominent, emphasizing enhanced traction and performance, and the tapering side-window line echoes that of the smaller Honda CRX, itself designed to appear more of a crossover than a weighty SUV. The side windows and black upper pillars give the MDX a longer, sleeker profile, and it rides lower on its 18-inch wheels too. From the rear, there is a distinct flavour of the Audi Q7, another well-received premium SUV.

Despite having electronically adjustable suspension honed on the dynamically demanding Nürburgring in Germany, the MDX, Acura insists, is above all a luxury car that offers a comfortable, quiet ride. In line with the demands of the pampered US consumer, the MDX provides a particularly sumptuous interior with soft leather seating for seven people, an elegant – rather than brash – dashboard and even a GPS-linked tri-zone auto climate control that automatically adjusts the temperature and fan speed according to the position of the sun.

Alfa Romeo 8C Competizione

Engine	4.7 V8
Power	336 kW (450 bhp) @ 7000 rpm
Torque	470 Nm (346 lb. ft.) @ 4750 rpm
Gearbox	6-speed manual
Installation	Front-engined/rear-wheel drive
Front suspension	Double wishbone
Rear suspension	Double wishbone
Brakes front/rear	Discs/discs
Front tyres	245/35R20
Rear tyres	285/35R20

Alfa Romeo's 8C Competizione is the production version of the concept of the same name first shown at Frankfurt in 2003 and featured in the *Car Design Yearbook 3*. In production form the exterior is almost identical to that of the concept; only a new wheel design and the addition of chrome trim around the curvaceous side windows count as visible changes.

Although the new car is shaped with a nod to the classic pre-war Alfa Romeo 8C cars and the '6C Competizione' sports coupé driven in 1950 by Fangio and Zanardi in the famous Mille Miglia race, it is claimed not to be a retro design. Yet it is certainly very successful in evoking the voluptuous aura of those classic models, despite the fact that, mechanically, it uses the bang up-to-date Maserati coupé platform as well as a more powerful version of its V8 engine.

It is without question a gorgeous design, full of delicious curves and with a version of Alfa's three-part front grille that is simple yet wonderfully attractive. The shoulder line along the side undulates up and down from front to rear, while the crease in the door draws a line linking the front wheel to the rear. The rear end truncates abruptly and is tipped by a subtle kick-up spoiler at the rear; it is all delightful.

The black-and-aluminium interior is the only design let-down. Parts do not fit flush with one another and the steering wheel looks like an aftermarket accessory. For a car that does not want to be viewed as retro, the interior should have been more modern in order to dispel any such thoughts.

This is a low-volume flagship model, similar in vein to BMW's Z8. It is aspirational, high in price and probably costly to build. But the buzz it has created around Alfa means that it is already a success.

CW 606GW

Audi A5

Engine	4.2 V8 (3.2 V6, and 2.7 and 3.0 diesel V6, also offered)
Power	354 kW (475 bhp) @ 7000 rpm
Torque	440 Nm (324 lb. ft.) @ 3500 rpm
Gearbox	6-speed manual
Installation	Front-engined/all-wheel drive
Front suspension	Double wishbone
Rear suspension	Multi-link
Brakes front/rear	Discs/discs
Front tyres	245/40R18
Rear tyres	245/40R18
Length	4635 mm (182.5 in.)
Width	1854 mm (73 in.)
Height	1369 mm (53.9 in.)
Wheelbase	2751 mm (108.3 in.)
Track front/rear	1594/1581 mm (62.8/62.2 in.)
Kerb weight	1630 kg (3593 lb.)
0–100 km/h (62 mph)	5.1 sec
Top speed	250 km/h (155 mph) limited
Fuel consumption	12.4 l/100 km (22.7 mpg)
CO_2 emissions	298 g/km

The new A5 is an intensely emotional introduction for Audi, for it marks the company's return to the high-performance coupé sector, which really put the brand on the international map with the original 1980s quattro. Based on the architecture that will also underpin the next-generation A4 and A6, the A5 will be a luxury-class competitor for the Mercedes CLK, if not quite for the S-Class-derived CL. In Audi's scheme the A5 – and the convertible to follow – will sit comfortably between the more sports-focused TT below and the ultra-exclusive R8 supercar at the very top; nevertheless, the A5 and its companion S5 mark an important shift in Audi design towards a more voluptuous, more curvaceous style.

The difference is most noticeable in the powerful stance and the strong shoulder lines. The sharp line that winds its way from the headlamps to the rear lamps is a very significant feature as it ties the front of the car to the rear and, by gently rising over the tightly fitted wheel arches, signifies a great deal of power potential from all four wheels.

The front, with its deep square grille, is classically Audi, but the attractive sharp-lipped rear is very untypical for the brand. The roofline extends far enough back to allow space for a decent-sized rear quarter-window and the promise of reasonable head room for the rear passengers. The front and rear bumpers are deliberately kept subtle so as to keep the design looking light and nimble.

The interior is finely detailed, trimmed with thin chrome rims around the instruments and switchgear. Grey and silver panels discreetly break up the inside for the sophisticated look that has for some time been typical of Audi designs. Overall, this is not a car filled with innovation, but it does mark a welcome return to more fluid and emotional design from Audi.

IN S 5020

Audi R8

Engine	4.2 V8
Power	309 kW (414 bhp) @ 7800 rpm
Torque	430 Nm (317 lb. ft.) @ 4500–6000 rpm
Gearbox	6-speed manual
Installation	Mid-engined/all-wheel drive
Front suspension	Double wishbone
Rear suspension	Double wishbone
Brakes front/rear	Discs/discs
Front tyres	235/30R19
Rear tyres	295/30R19
Length	4430 mm (174.4 in.)
Width	1900 mm (74.8 in.)
Height	1250 mm (49.2 in.)
Wheelbase	2650 mm (104.3 in.)
Kerb weight	1560 kg (3439 lb.)
0–100 km/h (62 mph)	4.6 sec
Top speed	301 km/h (187 mph)

The mid-engined R8 supercar is not just a whim from Audi: it is an integral element in the brand's steady, strategic climb towards the top of the market. Having already met and matched both Mercedes-Benz and BMW, Audi now clearly feels able to challenge Porsche.

The first warning shot was fired at the Frankfurt show in 2003 when Audi showed its Le Mans concept car. Based on Lamborghini architecture, the V10-powered prototype had an aluminium structure and a lightweight body that showed clear links to the successful TT but that took the Audi look into a more exotic dimension.

The production R8 remains almost completely faithful to the Le Mans concept, right down to many of the smallest details. The grown-up TT look is still there, though the only actual similarities are the sloping roofline and the treatment of the rear window. The optional Smart-like contrast panel behind the door (in dark carbon or polished aluminium, depending on body colour) is a clear differentiator, though opinions are divided on whether or not it breaks up the car's lengthwise movement too much. The Bugatti Veyron is perhaps the best-known car with two-tone paintwork, the intertwined colours of which run fore and aft.

Other shapes and details remain as elegant as on the Le Mans, with only the squared-up air intake corners distinguishing the roadgoing R8. The large slatted air intakes sit below the very striking headlamps, lined along the base with LED lighting.

The interior looks every bit the part, too, and strikes a good balance between high performance and usability. Particularly pleasing is the engine bay, where the 4.2-litre V8 nestles inside the beautifully finished aluminium spaceframe, all clearly visible through the rear screen.

Technically, the four-wheel-drive Audi is doubtless up there with the best. But whether it has the purity and balance to become a timeless design icon like the Porsche 911 is another matter.

IN R 805

R8
IN R 805

Audi TT

Engine	3.2 V6 (2.0 in-line 4 also offered)
Power	189 kW (250 bhp) @ 6300 rpm
Torque	320 Nm (236 lb. ft.) @ 2500–3000 rpm
Gearbox	6-speed manual
Installation	Front-engined/all-wheel drive
Front suspension	MacPherson strut
Rear suspension	Multi-link
Brakes front/rear	Discs/discs
Front tyres	245/45R17
Rear tyres	245/45R17
Length	4178 mm (164.5 in.)
Width	1842 mm (72.5 in.)
Height	1352 mm (53.2 in.)
Wheelbase	2468 mm (97.2 in.)
Track front/rear	1572/1558 mm (61.9/61.3 in.)
Kerb weight	1410 kg (3109 lb.)
0–100 km/h (62 mph)	5.9 sec
Top speed	250 km/h (155 mph) limited
Fuel consumption	8.3 l/100 km (34 mpg)
CO_2 emissions	247 g/km

Few design tasks can have been more daunting than that of creating the successor to the Audi TT. Launched in 1998, the original TT quickly became an icon both within the design community and among the broader public, and went on to win numerous awards.

Externally, the new car is unmistakably a TT, but is clearly also fatter, broader and beefier. The most marked difference can be seen in the side profile, where the longer overhangs and side windows betray the fact that it is 137 mm (5.4 in.) longer. The beltline is kept high, with a letter-box door window to peer out from. Since the original TT, Audi has standardized a new corporate face on all its other models: logically, the new TT has the tall grille that cuts right down into the bumper. There is more flair and energy at the front, with headlamps that now sweep up and out into the fenders, as is currently the trend.

In profile, the new TT retains the characteristic arched cant rail of the original, along with a similar wheel-hugging body. From the rear, the increased width can be seen: the lamps are wider, with the line created by their inner edge leading into the number-plate recess. The rear lacks the classic simplicity of the original – growing up has meant using more complex forms – but the new model is undeniably still a TT, and an attractive one at that.

The interior is of the usual high quality, mixing Alcantara and nappa leather with aluminium and chrome. The flat-bottomed sports steering wheel highlights the TT's sporty nature, while at the same time removing any nagging memory of the TT's being derived from the VW Golf.

Overall, while Audi's new TT may have lost the purity of form of the earlier car, it has evolved but has not lost its identity. As with the original, a roadster version is also offered.

IN TT 406

IN TT 160

Bentley Brooklands

Design	Dirk van Braeckel
Engine	6.75 V8
Power	395 kW (530 bhp)
Torque	1050 Nm (774 lb. ft.)
Gearbox	6-speed automatic
Installation	Front-engined/rear-wheel drive

While Rolls-Royce majors on the ultimate in big-limousine luxury, Bentley regards its heartland as high-performance big cars appealing to a more sporty type of driver. So the new Brooklands, in effect a coupé cousin to the Arnage limousine, is perhaps the ultimate expression of everything Bentley stands for.

Named after the famous UK racing circuit, the Brooklands is every inch a classic: it comes complete with a low roofline and a long waistline shoulder rising slightly over the rear wheels and sloping off at the rear. The view from the side is perhaps the best aspect of this car, in particular the large, forward-sloping C-pillar that gives such strong emphasis to the power emanating from the driven wheels at the rear.

Again true to its mission, the Brooklands looks a more British car than the top-selling and technically more sophisticated Continental GT coupé, with its ground-hugging Audi TT-like stance. By contrast the Brooklands, when viewed from the front or rear, can still look old-fashioned; this is entirely intentional, of course, and helps to evoke the discreet and powerful image desired by those customers who feel that the Continental is too raw and racy. Nevertheless, the relative lack of advance in these areas does illustrate the well-recognized problem of bringing progressiveness to the design of super-luxury vehicles.

The interior, in particular, remains true to the impeccable quality values established by Bentley early in the twentieth century, even though much contemporary technology lies behind the perfectly stitched leather, deep carpets and beautifully veneered panels.

This high-speed luxury travelling machine is emblematic of a modern age in which, with money, one can have almost anything. And though wider environmental concerns are increasing the social pressure on such examples of conspicuous consumption as this, the entire limited run of 550 Brooklands is sure to be highly prized among the world's élite buyers.

2000 TU

Bertone Barchetta

Design	Bertone
Brakes front/rear	Discs/discs
Front tyres	225/30R20
Rear tyres	225/30R20
Length	3585 mm (141.1 in.)
Width	1705 mm (67.1 in.)
Height	1090 mm (42.9 in.)
Wheelbase	2300 mm (90.6 in.)

In celebration of its ninety-fifth anniversary, Italian design house Bertone presented the Barchetta, a two-seater roadster based on the platform of the Fiat Panda. The car takes its references from mid-twentieth-century two-seater sports cars, in particular a one-off Fiat 500 with barchetta (an Italian term for a small open sports car) bodywork created by Nuccio Bertone in 1947.

The most striking thing about the Barchetta is the way its polished metal upper half appears to clamp down on the piano-black-and-glass lower structure, itself resting on the bright metal rocker panels. The proportions are doubly confused by the way in which the bodywork pinches inward behind the front wheels, enclosing the occupants tightly and creating a strong outward kick for the rear wheel arches. The rearward-opening scissor doors must be seen to be believed: they hinge through the silver-black interface around the rear wheel arch and are a Bertone patent.

The car looks very different with the doors swung shut: their heavy top rail makes for an unusually high waistline – further exacerbated by the glass panels that make up the lower half of the doors – and the visual mass of the front of the car becomes heavier still. The minuscule, pillarless windscreen heightens this unbalanced effect.

Inside, the tan hide seats are fixed in position; it is the simple white dashboard panel, with its four square instruments, that moves, along with the pedals and gear lever, to provide for different sizes of driver. The technology inside is kept simple but up to date, including an iPod docking port in the centre console.

For more than eighty years Bertone has had a strong relationship with Fiat. This revival of the barchetta idea is a sophisticated design that carries with it the originality worthy of a Bertone design statement: intriguing and appealing, even if it would be hopelessly impractical to manufacture and operate.

FIAT

BMW X5

Design	Chris Bangle
Engine	4.8 V8 petrol (3.0 in-line 6, and 3.0 in-line 6 diesel, also offered)
Power	265 kW (355 bhp) @ 6300 rpm
Torque	475 Nm (350 lb. ft.) @ 3400–3800 rpm
Gearbox	6-speed automatic
Installation	Front-engined/all-wheel drive
Front suspension	Double wishbone
Rear suspension	Multi-link
Brakes front/rear	Discs/discs
Front tyres	255/55R18
Rear tyres	255/55R18
Length	4854 mm (191.1 in.)
Width	1933 mm (76.1 in.)
Height	1766 mm (69.5 in.)
Wheelbase	2933 mm (115.5 in.)
Track front/rear	1644/1650 mm (64.7/65 in.)
Kerb weight	2245 kg (4949 lb.)
0–100 km/h (62 mph)	6.5 sec
Top speed	240 km/h (150 mph)
Fuel consumption	12.5 l/100 km (22.5 mpg)
CO_2 emissions	299 g/km

Anyone hoping to see a major step forward in design with the launch of the 2007 BMW X5 is destined to be disappointed. The new model does not mark the visual advance that its predecessor did in 1999; in fact, this is one of those second-generation models that are very hard to tell apart from the original.

Yet look more closely and it is clear that the proportion has changed: the wheels now appear less oversized in relation to the body mass, the waistline appears higher and the windows shallower, and the front is more rounded for a less aggressive impression. The fact that the new car does not look any larger is a tribute to its clever design, for the package has in fact been made 25 cm (10 in.) longer to cater for the third row of seats that, thanks to the success of the Volvo XC90, is fast becoming a market essential in the premium SUV segment.

Although the design at first appears somewhat bland, close study reveals subtle use of BMW's 'flame-surfacing' technique. It is this surfacing style that makes the new X5 look like a new car for 2007. Shallow U-shaped valleys in the door and bonnet panels rise up to form crisp peaks that emanate from the front grille and run along the side of the car through the door handles.

Design differences are easier to spot inside the vehicle, where the architecture is now much more carlike, and where the array of electronic chassis control systems marks the X5 out as perhaps the most technically sophisticated 4x4 on the market.

M VR 4342

Cadillac CTS

Design	John Manoogian
Engine	3.6 V6 (2.8 V6 also offered)
Power	224 kW (300 bhp) @ 6700 rpm
Torque	366 Nm (270 lb. ft.) @ 4000 rpm
Gearbox	6-speed automatic
Installation	Front-engined/all-wheel drive
Front suspension	SLA
Rear suspension	Multi-link
Brakes front/rear	Discs/discs
Front tyres	235/50R18
Rear tyres	235/50R18
Length	4766 mm (187.6 in.)
Width	1841 mm (72.5 in.)
Height	1472 mm (58 in.)
Wheelbase	2880 mm (113.4 in.)
Track front/rear	1575/1585 mm (62/62.4 in.)
Kerb weight	1860 kg (4100 lb.)

The new CTS is a stronger and more masculine evolution of the outgoing model. It retains much of the design character and proportion; however, the features are stronger and more intricate, giving it the air of a higher-quality car. In particular, the fine detail in the larger front grille that runs down in a tapered fashion through the bumper, the vent in the top of the fender, and the LED rear lamp that doubles up as a boot-lid spoiler all add to its higher-end feel. It is wider, too, which significantly improves its stance on the road and allows for an optional all-wheel-drive configuration; the greater tumblehome for the side glass helps the stance, too.

The rising crease line through the doors featuring on the old model has now been replaced with a horizontal shoulder line above the door handles, and a rising line lower down through the doors. This allows for a large swathe of uninterrupted panel through the centre of the doors, giving a contemporary and strong impression. More volume is added to the front of the car, but the tone is less aggressive.

North American cars, even Cadillacs, have frequently struggled to match the interior quality found on European luxury cars. The CTS is an important step in the right direction. The switch and instrument details are all carefully considered, and the overall feel of the interior is much softer than that of the masculine exterior. The integrated centre console comes in either a satin metallic finish for those who want a sports cabin, or sapele hardwood for those preferring the comfort option.

Sitting between the larger STS and the smaller BLS, the new CTS gives a professional, upmarket impression while keeping the classic pointed plan shape front and rear and a less blocky rendering of Cadillac's current chisel-edged look.

Chevrolet Malibu

Engine	3.6 V6 (2.4 in-line 4 also offered)
Power	188 kW (252 bhp)
Gearbox	6-speed auto
Installation	Front-engined/front-wheel drive
Front suspension	MacPherson strut
Rear suspension	Multi-link
Brakes front/rear	Discs/discs
Length	4872 mm (191.8 in.)
Width	1785 mm (70.3 in.)
Height	1451 mm (57.1 in.)
Wheelbase	2852 mm (112.3 in.)
Track front/rear	1514/1524 mm (59.6/60 in.)
Fuel consumption	10.7 l/100 km (26.4 mpg)

The Chevrolet Malibu has an unenviable role as the standard sedan for Middle America, the kind of car that is often considered to be forgettable, but which cannot afford to put a foot wrong or offend anyone's sensibilities. All too often, this has led to bland, faceless designs in this segment.

Fittingly, the new Malibu is a rather better-looking car than the one it replaces. The greater wheelbase and extended cabin are particularly helpful to the visual impression the car creates, and naturally increase the roominess inside, too. The new face of the Malibu has a much stronger brand identity, with the double-decker grille and V-shaped shoulder lines running back from the headlamps to the A-pillars. By comparison, the flat rear panel is a let-down. The proportion comes from the use of the front-wheel-drive GM Epsilon platform, which is also employed to underpin the Pontiac G6, the Saturn Aura and, in part, the Saab 9-3.

Designed to compete squarely with the new Honda Accord and the Toyota Camry, the Malibu has raised its game and is now presenting itself as a higher-quality car. The use of chrome is often perceived as the sign of quality; the Malibu comes with chrome-tipped exhaust pipes and even a rear number-plate surround trimmed in chrome. The simple but sculpted two-tone interior wraps itself around the occupants, with different trim and material combinations available to provide an element of individuality.

Whereas the outgoing model suffered at the hands of the strong and often slightly less bland opposition, this higher-quality Malibu is certainly an improvement, although not a totally convincing one. With a new Honda Accord on the way, the Malibu will have to fight hard to make its case.

Chevrolet Volt

Design	Anne Asensio
Engine	1.0 in-line 3 ethanol engine and electric motors
Power	140 kW (188 bhp)
Front suspension	MacPherson strut
Rear suspension	Torsion beam
Brakes front/rear	Discs/drums
Front tyres	195/55R21
Rear tyres	195/55R21
Length	4318 mm (170 in.)
Width	1791 mm (70.5 in.)
Height	1336 mm (52.6 in.)
Track front/rear	1626/1626 mm (64/64 in.)
0–100 km/h (62 mph)	8.5 sec
Top speed	193 km/h (120 mph)
Fuel consumption	1.6 l/100 km (150 mpg)

The Chevrolet Volt turns on its head the stereotype that American companies are unconcerned with energy-saving engineering. It is a high-profile showcase for GM's latest electric propulsion system, though technically speaking it is a plug-in hybrid.

The novelty of GM's E-Flex powertrain is that most everyday driving can be done using the big lithium-ion battery, charged up from the mains in six hours. But to extend the vehicle's range beyond the battery's 65-kilometre (40-mile) maximum, a small and highly efficient combustion engine kicks in to recharge the battery, giving an effective operating radius of up to 1030 kilometres (640 miles). The range-extending engine runs at its most efficient rpm, but never powers the wheels directly.

The show car's engine was designed to run on E85 ethanol, but GM makes the point that the recharging could be entrusted to a range of different power units: diesel, biofuel or even a fuel cell. The technology packed into the Volt is expected to be ready for mass production in 2010, the company says.

Not limited by the packaging constraints of conventional vehicles, the Volt's designers were able to push the large wheels right to the corners of the car. Designed in Michigan, the exterior employs polycarbonate resin to create the transparent roof, boot lid and side windows. The formability of this material when it is being manufactured allows for the windows to follow the sharp edge of the car's waistline, which runs the length of the car, although how the windows would drop into the doors is a mystery. The design is not overtly futuristic because GM wanted to make people aware that this technology is within reach. Yet its overall stance is athletic and interesting, and the more radical UK-inspired interior uses recyclable and lightweight materials and has a dual-mode instrument cluster that provides two levels of vehicle information.

The Volt shows that GM, although once derided for killing the EV-1 electric car, could yet turn the tables on its critics and introduce a much better model.

Chrysler Nassau

Design	Alan Barrington
Engine	6.1 V8
Power	317 kW (425 bhp) @ 6200 rpm
Torque	570 Nm (420 lb. ft.) @ 4800 rpm
Gearbox	5-speed automatic
Installation	Front-engined/rear-wheel drive
Front suspension	Short and long arm
Rear suspension	Multi-link
Brakes front/rear	Discs/discs
Front tyres	245/40R22
Rear tyres	245/40R22
Length	4981 mm (196.1 in.)
Width	1885 mm (74.2 in.)
Height	1496 mm (58.9 in.)
Wheelbase	3050 mm (120.1 in.)
Track front/rear	1621/1664 mm (63.8/65.5 in.)
Kerb weight	2041 kg (4500 lb.)
0–100 km/h (62 mph)	5 sec
Top speed	265 km/h (165 mph)

DaimlerChrysler defines the Nassau concept as a luxury four-door coupé, the currently fashionable genre kicked off by the Mercedes CLS-Class, which was launched at the Geneva Motor Show in 2004. The Nassau is based on the Chrysler 300's platform, so it carries across the hefty 6.1-litre Hemi V8, which Chrysler claims will propel this large sports car to 100 km/h (62 mph) in just five seconds.

Yet the concept presents something of a paradox, for it looks nothing like as long as its 5 metres (just over 16 ft.) when viewed directly from the side. Front and rear overhangs, especially, appear minimal, just like a compact hatchback. The explanation is twofold: the huge 22-inch wheels confuse one's sense of scale, making the bodywork look proportionately less massive, and the strong curvature in plan of both the front and the rear mean that the car is much longer than it at first appears.

When the car is viewed from the front, the lamps, grille and door mirrors are all elements with a geometric order to them, splaying out from the bold grille in the centre. At the rear there is a great degree of curvature in plan on the hatchback tailgate: the rear screen in particular is very curved, lending the Nassau a distinctive grown-up-small-car look that is unfamiliar in the luxury segment.

Prominent in profile are the fast windscreen and the cant rail, which drops down sharply through the rear door glass to meet the waistline. However, the raised centre line of the roof continues rearward at full head height, cleverly giving a spacious interior. Glass panels either side of this spine add further to the feeling of space.

The designers claim to have taken inspiration from mobile phones for the interior, particularly for the centre-console buttons and screen. Illuminated blue instruments sit in the dashboard and provide a fine contrast with their aluminium rims and the black fascia.

Chrysler Sebring

Engine	3.5 V6 (2.4 in-line 4 and 2.7 V6 also offered)
Power	175 kW (235 bhp) @ 6400 rpm
Torque	315 Nm (232 lb. ft.) @ 4000 rpm
Gearbox	6-speed automatic
Installation	Front-engined/front-wheel drive
Front suspension	MacPherson strut
Rear suspension	Multi-link
Brakes front/rear	Discs/discs
Front tyres	215/55R18
Rear tyres	215/55R18
Length	4842 mm (190.6 in.)
Width	1808 mm (71.2 in.)
Height	1498 mm (59 in.)
Wheelbase	2765 mm (108.9 in.)
Track front/rear	1569/1569 mm (61.8/61.8 in.)
Kerb weight	1599 kg (3525 lb.)
Fuel consumption	11.8 l/100 km (24 mpg)

Borrowing many of its styling cues from the Chrysler Airflite concept of 2003, the new Sebring makes a constructive contrast to its dowdy predecessor. Lending a sporty look are further design cues from the Crossfire coupé, including ribs along the bonnet and roof.

Perhaps reflecting DaimlerChrysler's European sales aspirations for the model, Chrysler chose July's London Motor Show for the world debut of the sedan version, while the stylish convertible derivative made its first public appearance at the LA show, perhaps with an eye on the California market.

For a mid-size saloon, the Sebring manages a distinctly individual style. At the front the characteristic egg-crate grille, quad headlamps and ribbed bonnet are unmistakably Chrysler. A strong forward-pointing stance is created mainly by the angled crease running through the doors but also by the rising waistline. It is also noticeable that the C-pillar comes down from the roof to merge with the waist behind the rear wheel: this gives larger rear doors, allowing better ingress for passengers as well as improved visibility to the rear. The rear doors actually include the front section of the wheel arch: this is an unusual arrangement as it can sometimes have the disadvantage of allowing dirt into the door surround area.

The interior shown has a high-class and very fashionable monochrome look, with a variety of textures and finishes to materials spanning a range of shades of grey. There are also cream and pebble-beige options on the higher-priced models. The car features a voice-activated entertainment system with new features for music, sound, videos and personalized picture displays; an important innovation is a built-in 20-gigabyte hard drive, enabling such accessories as a music jukebox.

The convertible is one of the few full-size four-seaters available. As well as having the novelty of the choice of three different finishes for its electric folding hard-top – metal, cloth or vinyl – it promises to be well priced, too.

SEBRING
CHRYSLER

Chrysler Town and Country

Design	Ralph Gilles
Engine	4.0 V6 (3.8 V6 and 3.3 flex-fuel V6 also offered)
Power	179 kW (240 bhp)
Torque	343 Nm (253 lb. ft.)
Gearbox	6-speed automatic
Installation	Front-engined/all-wheel drive
Brakes front/rear	Discs/discs
Length	5142 mm (202.4 in.)
Width	1953 mm (76.9 in.)
Height	1750 mm (68.9 in.)
Wheelbase	3078 mm (121.2 in.)
Track front/rear	1651/1646 mm (65/64.8 in.)

The Chrysler Town and Country shares its design with the Dodge Grand Caravan, the mainstay of the American soccer-mom brigade, and the cars are to all intents and purposes identical. Both appeared in new editions at the 2007 Detroit show, by an unfortunate coincidence just as the US market for large minivans was beginning to weaken.

The model's makeover preserves elements from the look that has become familiar over several preceding generations; this latest iteration has a bulkier and more upright design, doing little to soften its impression. Most noticeably, the bonnet is now a separate element, more horizontal, like that of a conventional car, rather than continuing the slope of the windscreen. The new car has slightly more pronounced wheel arches and crisp feature lines running from the front fender back above the door handles.

The interior, arguably the most important aspect of any people-carrier, goes one better than previous generations with a new 'swivel and go' system for the second-row seats that allows them to rotate 180 degrees to face the power-folding third row. Further features include such innovations as an industry-first integrated child booster seat, a rear reversing camera, a rear-view interior conversation mirror and a large upright centre console containing the navigation system and an analogue clock sunk into an aluminium plate. Ambient halo lighting and pinpoint reading lamps for all passengers are another Chrysler–Dodge boast, as is a sliding and removable centre console.

Since it created the sector in 1983 the Chrysler Group has sold more than eleven million minivans. This vehicle will be sold in Europe under the Chrysler brand as the Grand Caravan, and Chrysler's plant will produce a version for Volkswagen, to be styled, VW stresses, very differently inside and out.

Citroën C4 Picasso

Engine	2.0 in-line 4 (1.8 also offered)
Power	103 kW (138 bhp) @ 6000 rpm
Torque	200 Nm (147 lb. ft.) @ 4000 rpm
Gearbox	6-speed automated manual
Installation	Front-engined/front-wheel drive
Front suspension	MacPherson strut
Rear suspension	Trailing arm
Brakes front/rear	Discs/discs
Front tyres	215/50R17
Rear tyres	215/50R17
Length	4590 mm (180.7 in.)
Width	1830 mm (72 in.)
Height	1680 mm (66.1 in.)
Wheelbase	2728 mm (107.4 in.)
Track front/rear	1505/1539 mm (59.3/60.6 in.)
Kerb weight	1560 kg (3439 lb.)
0–100 km/h (62 mph)	11.5 sec
Top speed	195 km/h (121 mph)
Fuel consumption	8 l/100 km (35.3 mpg)
CO_2 emissions	190 g/km

The design world has a love–hate relationship with people-carriers. Yes, they open up new frontiers of practicality and space efficiency, but they are a notoriously difficult vehicle type to make look exciting or even halfway interesting. For too long they have been large, bland boxes and little more.

The Citroën C4 Picasso may well change all that. Though its high style seems a long way from that of the streamlined C4 hatchback, it picks up themes from the recent C-SportLounge and C-AirLounge concepts and succeeds in bringing poise and creative tension to this unfashionable format. In terms of size it sits between the Xsara Picasso and the big C8, making it usefully bigger than the segment-defining Renault Scenic but without the bulk of a full-size MPV.

Citroën has made the Picasso interesting through careful attention not just to its overall proportions but also to its detailing. The plunging window line creates a sense of forward movement, while the arched DLO and deep side glass project a feeling of space; angular corners, most noticeable around the windows and rear lamps, add sharpness, while the slim, split A-pillar is a neat solution to this perennial MPV problem and gives the upper architecture a much lighter feel, a feature it has taken from the C-SportLounge concept.

The interior is of an equally high standard, with a smooth, sweeping dashboard, a floor unencumbered by a gearlever (the principal transmission is a six-speed automated manual with a stalk-mounted lever) and a fixed-hub steering wheel, which helps to keep the dashboard free of almost all switches. An Espace-style split heating system sees controls at either end of the dash, and also new are the scented air freshener, lane departure warning system and a newly developed parking-space gap sensor.

This is an exceptionally well-resolved seven-seater and, more recently, Citroën has added a five-seater with a differentiated style. At last Renault may have met its match in design.

"sculpte par le vent"
rangement ouvert
pad elasto
GONZALEZ

Citroën C-Crosser

Engine	2.2 in-line 4 diesel
Power	116 kW (156 bhp) @ 4000 rpm
Torque	380 Nm (280 lb. ft.) @ 2000 rpm
Gearbox	6-speed manual
Installation	Front-engined/all-wheel drive
Front suspension	MacPherson strut
Rear suspension	Multi-link
Brakes front/rear	Discs/discs
Length	4640 mm (182.7 in.)
Width	1810 mm (71.3 in.)
Height	1710 mm (67.3 in.)
Wheelbase	2670 mm (105.1 in.)
Fuel consumption	7.3 l/100 km (38.7 mpg)
CO_2 emissions	194 g/km

PSA–Peugeot Citroën has at last bowed to consumer pressure and launched an SUV, courtesy of a co-operative deal with Mitsubishi of Japan. Citroën's C-Crosser and Peugeot's 4007 are mechanically identical, both based on the Mitsubishi Outlander, itself built on the new platform shared with Chrysler.

Though built in Japan, both French-branded versions differ from Mitsubishi's own product in that they use PSA's well-respected 2.2-litre diesel engine, which can run on up to 30 per cent biofuel. Important body style differences distinguish all three versions, especially at the front.

Citroën's designers have managed to create a car with good looks, in particular the attractive chevron grille and lamps that draw the eye up; combined with the black lower grille, these visual devices help to disguise the inevitable height of an SUV front end.

There is no low-range transfer gearbox available for either model, which confirms that this range has no intention of expanding beyond the mild off-roading capabilities of a market-pleasing crossover. Instead, this is a practical family car: it seats seven, if so desired, there is a big boot, and the tailgate is hinged top and bottom and splits in the centre for convenient loading. The instrumentation and features are minimal compared with the car's more lavish and generally more expensive rivals.

Mitsubishi's petrol engines may be available as an option some time after launch, but the European market is heavily skewed towards diesel and in any case Citroën – like Peugeot – has tweaked its version more towards comfort than towards sportiness and driver feel.

CITROËN
601 DYP 78
C-CROSSER

601 DYP 78

Citroën C-Métisse

Engine	V6 diesel hybrid with electric motors driving the rear wheels
Power	160 kW + 30 kW (208 bhp + 40 bhp)
Gearbox	6-speed automatic
Installation	Front-engined/all-wheel drive
Front suspension	Double wishbone
Rear suspension	Double wishbone
Brakes front/rear	Discs/discs
Front tyres	255/40R20
Rear tyres	255/40R20
Length	4740 mm (186.6 in.)
Width	2000 mm (78.7 in.)
Height	1240 mm (48.8 in.)
Wheelbase	3000 mm (118.1 in.)
0–100 km/h (62 mph)	6.2 sec
Top speed	250 km/h (155 mph)
Fuel consumption	6.5 l/100 km (43.5 mpg)
CO_2 emissions	174 g/km

After many years of unexciting middle-of-the-road models, Citroën is in the midst of a full-blown design revival, with a number of exciting-looking cars already having reached the market. Each was preceded by an ambitious show concept: the 2002 C-Airdream, for example, in many ways previewed the production C4 hatchback.

Although very conceptual at this stage, the imposing C-Métisse, a racy and low-slung four-seater grand tourer, could be a plausible stepping stone to a future four-door coupé, Peugeot Citroën having publicly stated that it intends to enter six new market segments.

Beautifully evocative, the C-Métisse is a striking prospect: the visual mass is located far back over the rear wheels, giving a sense that the car is propelled from the rear, even though the hybrid diesel engine drives the front and electric motors power the rear wheels.

The steeply raked wraparound windscreen blends into the long letter-box side windows that make the car look extremely racy. At the front, irregular shapes sit side-by-side with large sculpted air intakes and a central grille in the bumper, suggestive of the power that lies behind its face. Angles at the top of the front fenders and in the headlamps and the rear lamps echo the angle made by the Citroën chevron.

Conceiving the interior for a futuristic car such as this is a dream job for a designer. A stark mix of white and black leather has been used, with aluminium for parts that interact with the driver, such as the pedals, gearshift and instruments; intriguingly, the head restraints come down from the roof.

Though this is unquestionably a spectacular car, it does come with a few rather major technical obstacles, not least of which is the immense complexity of the doors. These would need to be addressed before it could be considered for manufacture.

CITROEN

Daihatsu D-Compact X-Over

Design	Italdesign
Engine	1.5 in-line 4
Power	96 kW (129 bhp) @ 7000 rpm
Torque	140 Nm (103 lb. ft.) @ 4400 rpm
Gearbox	4-speed automatic
Installation	Front-engined/front-wheel drive
Front suspension	MacPherson strut
Rear suspension	Torsion beam
Brakes front/rear	Discs/discs
Front tyres	205/50R17
Rear tyres	205/50R17
Length	3750 mm (147.6 in.)
Width	1695 mm (66.7 in.)
Height	1575 mm (62 in.)
Wheelbase	2440 mm (96.1 in.)
Kerb weight	1020 kg (2249 lb.)

The D-Compact X-Over is one of Daihatsu's most recent concepts, yet despite its name it has nothing to do with the current fashion for crossovers (high-riding hatchback-estates with a touch of SUV thrown in). Instead, the X-Over expands the Daihatsu D-Compact family in a different direction: that of a mainstream B-segment family hatchback.

What makes it different, however, is its design pedigree. The model has been conceived in conjunction with Italdesign, the organization responsible for many of the industry's most memorable vehicles. The difference shows in the fresh clarity of the X-Over's lines, especially when the car is viewed from the rear. Using the starting point of the same platform as the new Materia, the styling brief for the X-Over was to follow a theme described as 'natural harmony'. Shown in a natural green colour – a shade not often seen on show cars – the X-Over has a simplicity and purity of style reminiscent of such cars as the new Fiat Punto.

The windscreen runs right over the roof to flood the cabin with light and bring the outdoors indoors, as is appropriate to the theme. Large blistered (bulging) wheel arches and the low, dark grille in the front bumper visually help to ground the car and give it a dynamic look. The waistline kicks up at the back of the rear door, while a small stepped feature in that door adds a sporting touch by evoking forward motion: it also helps to break up what would otherwise be the dull lozenge shape of the door.

A single large piece of glass is used to form the tailgate, developing the theme followed by the roof. The proportion of glass compared with that of the coloured body panels is high: Italdesign is likely to have proposed this to make sure the X-Over is seen as an adventurous new car that will draw people in to investigate it more closely.

D-COMPACT
X-over

D-COMPACT
X-over

Daihatsu Materia

Engine	1.5 in-line 4
Power	76 kW (102 bhp) @ 6000 rpm
Torque	132 Nm (97 lb. ft.) @ 4400 rpm
Gearbox	5-speed manual
Installation	Front-engined/front-wheel drive
Front suspension	MacPherson strut
Rear suspension	Torsion beam
Brakes front/rear	Discs/drums
Front tyres	185/55R15
Rear tyres	185/55R15
Length	3800 mm (149.6 in.)
Width	1690 mm (66.5 in.)
Height	1635 mm (64.4 in.)
Wheelbase	2540 mm (100 in.)
Track front/rear	1470/1465 mm (57.9/57.7 in.)
Kerb weight	1035 kg (2282 lb.)
Fuel consumption	7.2 l/100 km (39 mpg)
CO_2 emissions	169 g/km

The quirky Daihatsu Materia is a versatile supermini-sized tall hatchback in the current vogue. Indeed, it is based on the D-Compact Wagon, a model originally shown six months earlier at the Geneva Motor Show in 2006. In its role it sits midway between a standard compact hatchback, such as a Renault Clio, and a more utilitarian compact wagon, Renault's Kangoo being the obvious example.

Extending the Renault parallel, the Materia is positioned against the Modus, yet, thanks to the boxy shape created by its rather upright windscreen and flat, horizontal bonnet, its proportions are much more unusual. This serves to distinguish it from not just the Modus but also the Nissan Note, the Opel Meriva and the Fiat Idea, all of which have adopted the monovolume style that sees the grille, bonnet, windscreen and roof blend into a single, seamless profile.

The Materia is designed to stand out from the crowd and generate love–hate reactions. As a small and relatively unknown brand – in Europe at least – Daihatsu needs to make an extra-special effort to put itself on to the automotive map and into consumers' consciousness. Such a product as this will help, though from an image point of view Daihatsu would be well advised to avoid becoming typecast as the company that makes purely oddball designs.

Powered by a 1.5-litre petrol engine and with plenty of space inside, this is clearly a model that will be practical and economical, and eminently suitable as family transport. From a design perspective, the inset square rear side window gives it a sense of solidity like that of a security van, while at the rear the lamps are set outside smiling silver trim that sits just above the bumper. This makes a clear statement that the Materia is targeting those with a non-conformist streak, even if its interior is by comparison disappointingly conventional.

MATERIA

MATERIA

MATERIA

Dodge Avenger

Design	Trevor Creed
Engine	3.5 V6 (2.4 and 2.7 in-line 4 also offered)
Power	175 kW (235 bhp) @ 6400 rpm
Torque	315 Nm (232 lb. ft.) @ 4000 rpm
Gearbox	6-speed automatic
Installation	Front-engined/all-wheel drive
Front suspension	MacPherson strut
Rear suspension	Multi-link
Brakes front/rear	Discs/discs
Front tyres	215/60R17
Rear tyres	215/60R17
Length	4848 mm (190.8 in.)
Width	1824 mm (71.8 in.)
Height	1496 mm (58.9 in.)
Wheelbase	2765 mm (108.9 in.)
Track front/rear	1569/1569 mm (61.8/61.8 in.)
Kerb weight	1696 kg (3739 lb.)
Fuel consumption	11.8 l/100 km (24 mpg)

Replacing the long-serving Stratus, the new Dodge Avenger saloon shares DaimlerChrysler's GS platform with the Chrysler Sebring, launched a matter of months earlier. What are termed mid-size saloons in North America account for almost a third of the overall passenger-car market in the US, and the proportion is set to grow further as the level of fuel prices inches upward and some awareness of climate-change issues begins to enter the American consciousness.

The Avenger is a bold and powerful-looking vehicle that immediately stands out on a crowded city street. When viewed from the side, the raised bonnet and boot are noticeable in relation to the lower window line, and the muscular raised rear haunches are particularly pronounced. The bonnet overhangs the characteristic Dodge grille, while the lower bumper has a small spoiler to help split the airflow between air to cool the radiator and air that is to run undisturbed under the car. A lip mounted on the trailing edge of the high rear deck helps to tidy the car's wake and reduce drag.

From the front, the wide grille with its signature cross-hair chrome bars and central Dodge logo is aggressive and unmistakably Dodge, as are the headlamps with their reflectors peering from underneath the bonnet. The muscle-car rear end ensures that this theme is followed through.

Interior innovations that appear to be aimed at US owners include a chilled storage compartment in the top of the instrument panel, holding up to four drinks cans, as well as a heated/cooled cupholder that keeps cold beverages at 1.6°C (35°F), and heats hot beverages to 60°C (140°F). There is voice-activation for music, sound, movies and personalized picture displays, while voice memo recording is available using the microphone integrated into the rear-view mirror.

Dodge Demon

Design	Jae Chung and Dan Zimmermann
Engine	2.4 in-line 4
Power	128 kW (172 bhp) @ 6000 rpm
Torque	224 Nm (165 lb. ft.) @ 4400 rpm
Gearbox	6-speed manual
Installation	Front-engined/rear-wheel drive
Brakes front/rear	Discs/discs
Length	3974 mm (156.5 in.)
Width	1736 mm (68.3 in.)
Height	1315 mm (51.8 in.)
Wheelbase	2429 mm (95.6 in.)
Kerb weight	1179 kg (2599 lb.)

With an eye to gaining a firm foothold in the European market and at the same time defending the brand in North America against the Mazda MX-5, the Saturn Sky and the Pontiac Solstice, Dodge's product planners decreed a back-to-basics two-seater sports car as a means of generating the requisite excitement. The result is the front-engined, rear-drive Demon concept, launched at the Geneva show to general approval.

First impressions are of a scaled-down Dodge Viper (front) meeting a Mazda MX-5 (centre section) with a bit of Audi TT thrown in at the very rear. The Demon's most distinctive feature, apart from its almost truck-sized Dodge-typical cross-hair grille, is its rear fender design, the swollen rear arches rising up and sweeping inward to meet the falling rear deckline in a not altogether dignified big-butt look. The TT connection comes in the shape and location of the rear lights and the way that the car's rear is very curved when viewed in plan.

This means that the car looks quite bulky when viewed from the front (or rear) three-quarters, but that it looks sporty, agile and compact, with a wheel at each corner, when seen side-on. From the side there are many design elements intended to communicate dynamism, from the unusual shape of the front wheel arch to the angled air scoop in the leading edge of the bulging rear wheel arch. The boot lid rises up at an angle where it meets the seat headrests, creating a cosy cabin right in the centre of the car and resolving a notoriously difficult area in roadster design.

Inside, the Demon smacks of pure 1960s classic sports cars, with black leather and metal – and quite a lot of plastic – combining to form a simple and effective driver-focused interior that spells driving fun rather than luxury or sophistication.

EDAG LUV

Design	Johannes Barckmann
Engine	6.1 V8
Power	345 kW (462 bhp)
Installation	Front-engined/all-wheel drive
Brakes front/rear	Discs/discs
Front tyres	295/35ZR22
Rear tyres	295/35ZR22
0–100 km/h (62 mph)	6.2 sec
Top speed	250 km/h (155 mph)

As the world's car manufacturers focus more on the building and marketing of cars and rely more heavily on the supplier community to engineer and supply not only parts but also whole vehicle systems, such companies as automotive design consultancy EDAG are eager to show off their knowledge of vehicle integration. In this complex process, a specification for a vehicle (or part of a vehicle) is obtained, the components are sourced, and everything is made to work smoothly together, ensuring that it can all be manufactured with no snags. Such a task might be anything from adding a new gearbox option to creating a complete vehicle.

The LUV, which EDAG describes somewhat paradoxically as a Luxury Utility Vehicle, is a kind of tough utility truck with a huge 6.1-litre V8 powerplant. Based on a Mercedes platform, the LUV uses know-how from the tuning house Brabus on the chassis and interior design, with EDAG co-ordinating the whole process. For the LUV the designers chose a maritime look, perhaps brought back into vogue by the Rolls-Royce Phantom convertible. Oiled teak features not only on the floor inside but also on the bonnet; the interior is trimmed in striking black-and-white leather.

But all this is not the point. The challenge is that the design has deliberately been driven by engineering and manufacturing feasibility departments, and the solution has been to develop a modular concept that allows different derivatives of the LUV to be made while still sharing a majority of parts between models. For example, the pickup module can be exchanged for a convertible or coupé module simply by releasing a catch mechanism integrated into the body. EDAG hopes its solutions will be attractive to car manufacturers in creating cost-effective niche vehicles.

Fiat Bravo

Design	Frank Stephenson
Engine	1.9 in-line 4 diesel (1.4 petrol also offered)
Power	110 kW (148 bhp) @ 4000 rpm
Torque	305 Nm (225 lb. ft.) @ 2000 rpm
Gearbox	6-speed manual
Installation	Front-engined/front-wheel drive
Front suspension	MacPherson strut
Rear suspension	Torsion beam
Brakes front/rear	Discs/discs
Front tyres	205/55R16
Rear tyres	205/55R16
Length	4336 mm (170.7 in.)
Width	1792 mm (70.6 in.)
Height	1498 mm (59 in.)
Wheelbase	2600 mm (102.4 in.)
Track front/rear	1538/1532 mm (60.6/60.3 in.)
Kerb weight	1360 kg (2998 lb.)
0–100 km/h (62 mph)	9.0 sec
Top speed	209 km/h (130 mph)
Fuel consumption	5.6 l/100 km (50.4 mpg)
CO_2 emissions	149 g/km

After the remarkable turnaround in Fiat's fortunes brought about by the huge success of the Grande Punto supermini following its launch in 2005, all eyes have been on the Italian company to see whether it can repeat the coup with the replacement for the larger Stilo, a solidly engineered hatchback that had been disastrously unsuccessful in competing against the Volkswagen Golf, the Ford Focus and other lower–medium models. And the unanimous opinion after the unveiling of the Bravo at the Geneva show was that Fiat could indeed pull it off again.

Superficially, the Bravo looks like a scaled-up Grande Punto: eye-catchingly attractive, with a streamlined, racy nose rising to a high, almost coupé-like tail. It is all very well resolved, too, which is remarkable considering that the design, developed by Frank Stephenson, was brought to market in super-quick time. (For more details on the work and career of Frank Stephenson, see the Profiles of Key Designers, page 264.) The only slightly awkward angle is from the rear three-quarter, when the rear corner can seem a touch heavy as the waistline rises to provide the distinctive sporty profile; in every other respect the Bravo is a beautifully elegant design.

The interior is more restrained and less distinctive, but well executed and pleasantly driver-focused nevertheless: as with the Punto, it marks a useful step up in quality and finish.

The Bravo is the first model to carry the new and much more confident Fiat logo, with its vertically elongated letters standing out against a ruby-red background. It is a fitting metaphor for a company that has rediscovered its Italian design flair: the friendly Bravo face, with its big headlamps and small grille, will sit comfortably alongside the much-anticipated new Fiat 500, launched in July, as well as the already successful Grande Punto.

FIAT
DA 284 DF

FIAT
Bravo

Fioravanti Thalia

Fioravanti, based in Turin and headed by former Pininfarina designer and ex-chief of Fiat's design centre Leonardo Fioravanti, is in Italian design terms quite new, moving from designing Tokyo golf courses to car design in 1991.

In classical mythology, Thalia was one of the fabled muses of poetry and comedy; for Fioravanti, the Thalia concept is a study of how to manage and disguise proportion, in its case an unusual double roofline layout: If this car were painted in a solid colour it would look overly heavy at the rear, and more like a clumsy estate car. Yet by the skilful use of contrasting colours and long curved lines the Thalia is actually made to look surprisingly dynamic.

The rear roof is designed as a glass dome to allow light to cover the rear-seat passengers, who are seated high up and so effectively get their own windscreen to look through. The decision to raise the rear passengers by about 30 cm (12 in.) gives plenty of space under the seats to package whatever power system is required; Fioravanti gives little guidance in this respect, and the exterior-only model shown at Geneva gave no clue. Other planned innovations include replacing the side mirrors with video cameras fitted on the roof spoiler. The reversing light, rear fog light and rear screen wiper are all integrated in the distinctive asymmetrical element on the tailgate that also includes the tailgate handle.

This Fioravanti concept is designed to show the car world how the company is able to come up with original design thoughts. There is some clever thinking in the packaging of the Thalia, ideas that may well come to be explored further by other designers in years to come.

Design	Leonardo Fioravanti
Length	4675 mm (184 in.)
Width	1880 mm (74 in.)
Height	1440 mm (56.7 in.)
Wheelbase	2900 mm (114.2 in.)

Ford Airstream

The Ford Airstream concept is intended to capture an exciting sense of adventure and discovery in its homage to the Airstream caravan trailers of past decades. Since 1931, when Airstream's first trailers were made from aluminium using aircraft construction methods, this elegant and iconic design has been emblematic of America's desire for travel.

Ford's new concept uses conventional car-construction techniques, with only a few superficial rivets acting as a nod to aircraft methods. Highly innovative, however, is the series hybrid powertrain, where a hydrogen-fed fuel cell feeds current to the battery, which then powers the wheels via an electric motor.

Though the proportions of the overall vehicle are very conventional, its graphic language is just the opposite: the T-shaped slab of chrome on the front is echoed by a similar T-shaped opening for the rear door, and the side windows are skewed rearward for dramatic graphic effect.

The three large doors hinged at the roof provide open access to the rich red interior, something that contrasts well with the bright aluminium exterior panels. Retro design elements are particularly noticeable in the rounded corners of all the windows. Pod-shaped front seats that can rotate like captain's chairs are inspired by furniture design of the 1960s, as are the side-facing rear benches. The dashboard, however, is bang up-to-date, with a floating instrument panel featuring touch-sensitive controls and a multi-function display for the driver. In the rear the focal point is a huge cylindrical screen that creates ambient mood settings, which include a lava lamp and even a virtual fire.

Ford's tie-up with Airstream could prove to be a shrewd commercial move if it can be translated into a production reality. Both brands are important American icons and, with Ford seeing crossovers as a key plank in its future product planning, the coolness associated with Airstream could be just the thing to help Ford secure a good slice of business from the new generation of idealistic baby-boomers now approaching retirement.

Design	Peter Horbury
Engine	Hybrid: hydrogen fuel cell/electric
Front suspension	Double wishbone
Rear suspension	Multi-link
Brakes front/rear	Discs/discs
Length	4699 mm (185 in.)
Width	2004 mm (78.9 in.)
Height	1793 mm (70.6 in.)
Wheelbase	3198 mm (125.9 in.)
Track front/rear	1717/1758 mm (67.6/69.2 in.)
Fuel consumption	5.7 l/100 km (41 mpg)

AIRSTREAM
Ford
MICHELIN
AIRSTREAM
Ford

Ford Focus

Engine	2.0 in-line 4
Gearbox	5-speed manual
Installation	Front-engined/front-wheel drive
Front suspension	MacPherson strut
Rear suspension	Multi-link
Brakes front/rear	Discs/drums
Length	4470 mm (176 in.)
Width	1722 mm (67.8 in.)
Height	1486 mm (58.5 in.)
Wheelbase	2614 mm (102.9 in.)
Track front/rear	1478/1468 mm (58.2/57.8 in.)

The European Ford Focus was completely redesigned in 2004. When it came to overhauling the North American version for 2008, however, Ford's managers decided not on an all-new vehicle but on an extensive overhaul of the existing platform, which dates back to 1998.

The new car has been carefully remodelled to give it a slightly more upmarket feel. A wide chrome-bar grille has been added to the front, visually widening the impression; this band flows into the headlamps, which are also very different from the originals of 1998. A recessed door area emanates from the vent at the back of the front fender, gradually rising through the doors until it reaches the rear lamps. A very nicely orchestrated modern touch is the way in which Ford's designers have encircled the red lenses of the rear lamps with clear glass on three sides. The flared wheel arches that are such a distinctive feature of the Focus remain, but they now flow seamlessly into the panels rather than looking like separate bolt-ons.

The model range has also changed, with the estate version no longer available and its place being taken by a two-door coupé. The more sporting coupé has a higher-seeming waistline than the saloon, and uses blue illumination of switches throughout the interior to blend with the aluminium T-panel used for the centre console. There is also an ambient lighting option – as on the new Mini from BMW – to give a choice of seven colours to light the interior.

The Focus was first introduced back in 1998 and at that time it represented a genuinely cutting-edge design. Sadly, the new US version of the car is nowhere near as radical a design statement. Not only is Ford slackening its pace but also the competition is catching up.

Ford Interceptor

Design	David Woodhouse and Kris Tomasson
Engine	5.0 V8
Gearbox	6-speed manual
Installation	Front-engined/rear-wheel drive
Front suspension	Double wishbone
Rear suspension	3-link design with panhard rod
Brakes front/rear	Discs/discs
Front tyres	22 inch
Rear tyres	22 inch
Length	5121 mm (201.6 in.)
Width	1941 mm (76.4 in.)
Height	1392 mm (54.8 in.)
Wheelbase	3068 mm (120.8 in.)
Track front/rear	1689/1722 mm (66.5/67.8 in.)

The long, wide and low silhouette of the Interceptor is very similar to the Ford 427 concept first shown in 2003, although the Interceptor is 35 cm (14 in.) longer and slightly taller. Yet though the new concept is a clear evolution of the 427, it has drawn more headlines for its brave bid to serve as a four-door Mustang, and for its attempt to apply to a passenger car the brutish three-chrome-bar frontal styling of last year's gargantuan Super Chief concept truck.

The Interceptor concept is indeed based on a stretched Mustang platform, but comes across as more massive in every way: it has a long wheelbase, huge wheels, deep sides and a rearward-biased cabin for an authentic low-rider look. The extended power bulge on the bonnet, the high waistline and the narrow glass areas accentuate this mean, aggressive impression. Peter Horbury, Ford's executive director of design, is quoted as saying that this car shows people what 'modern muscle' is all about. It seems indeed that this is about muscular proportions and features, but with soft rounded edges that hint at a new social responsibility.

Decidedly less contemporary is the recurring theme of squares with rounded-off corners. Familiar from the 1960s, this shape occurs in the rear lights and the low-set exhausts, as well as many dozens of times in the interior. The seats are inset with chrome-rimmed rounded-off rectangles, the gearshift, air vents and instruments pick up the same theme and, unforgivably, the steering wheel is also a square with rounded corners. All too often in the past, designers have attempted to impose a square steering wheel on car buyers, and each time the response from customers has been one of derision and disgust; it is surprising that car companies have not yet got the message.

Ford

Ford Iosis X

Design	Stefan Lamm and Nikolaus Vidakovic
Engine	Four-cylinder
Installation	Front-engined/all-wheel drive

This is not the first time that Ford has employed the Iosis name: the brand was first exposed at the Frankfurt show in September 2005 as a saloon, a theme that was re-presented a year later as the production-ready next-generation Mondeo. Now with an 'X' added for good measure, the Iosis has become a crossover concept. The link is clear: this is a 'soft' 4x4 based on the car platform that underpins the new Mondeo. Ford styles it a CUV: a Crossover Utility Vehicle.

In the metal, the design of the Iosis X comes across as brash, brutish and quite aggressive. The shaping of the car follows what Ford terms its 'kinetic' design language, consisting of features that generate visual energy around the chassis and powertrain aspects of the car. Huge tyres, deep sides and a shallow DLO dominate the look of the vehicle, with the side view showing a marked forward lean.

Large air intakes in the front bumper, arrowhead features in the headlamps, sculpted wheel arches and aluminium-contrast lower-body trim accentuate a complex, muscular look that would have to be toned down for broad commercial appeal, as would the vast wheels and high-mounted door mirrors. In addition, proper door handles would be needed in some form, together with a step to enable passengers to access the cabin.

The interior is futuristic, with design ideas said to be taken from modern helicopter-console layout. Long, streamlined shapes and the strong band of orange down the centre console create an almost sports-car-like feel, despite the vehicle's high ground clearance. The use of orange colour within the tyre tread is novel.

Production versions of the Iosis X are expected by the first quarter of 2008; as with the Mondeo, considerable cost savings are expected through the use of an enlarged Focus engineering platform.

Ford
IosisX

Ford Mondeo

Design	Martin Smith
Engine	2.5 in-line 5 (1.6, 2.0, 2.3 in-line 4, and 1.8 and 2.0 in-line 4 diesel, also offered)
Power	163 kW (218 bhp) @ 5000 rpm
Torque	320 Nm (236 lb. ft.) @ 4800 rpm
Gearbox	6-speed manual
Installation	Front-engined/front-wheel drive
Front suspension	MacPherson strut
Rear suspension	Multi-link
Brakes front/rear	Discs/discs
Front tyres	235/45R17
Rear tyres	235/45R17
Length	4844 mm (190.7 in.)
Width	1886 mm (74.3 in.)
Height	1500 mm (59 in.)
Wheelbase	2850 mm (112.2 in.)
Track front/rear	1589/1605 mm (62.6/63.2 in.)
Kerb weight	1492 kg (3289 lb.)
0–100 km/h (62 mph)	7.7 sec
Top speed	240 km/h (150 mph)
Fuel consumption	9.3 l/100 km (30.4 mpg)
CO_2 emissions	222 g/km

The new Mondeo is important both commercially and psychologically to Ford, for it represents a leading volume carmaker's response to the growing dominance of premium and near-premium brands in the European market for medium–large saloons, which are often the mainstay of company car fleets. The message is a clear one: Ford's so-called kinetic design language gives the Mondeo a much classier look and appeal than the now tired-looking model it replaces. But will this be enough to prevent buyers migrating to Volkswagen, BMW and Audi?

Proportionally, the car is extremely well balanced, confidently communicating a certain sportiness, solidity and interior spaciousness through dynamic lines, a well-resolved and tight wheel-arch-to-wheel relationship, and a long cabin with rear quarter-light. The prominent shoulder line on the bonnet leads down to the large headlamp, the shape of which clearly proclaims the new Mondeo as a progressive and dynamic car. The doors and bonnet are made up of taut surfaces broken up with strong feature lines that fade in and out: the doors, for example, have a linear depression at the bottom that gradually fades into the rear wheel arch. The new model's sophistication comes from careful attention to a number of details, such as the waistline kicking up at the rear quarter-window to parallel the line of the rear lamps.

The sporty interior design carries through some of the exterior themes, and in terms of technology the new car is important in introducing adaptive cruise control and hill launch assist to the volume segment. This is also the first car to feature Ford's Easyfuel system, which guarantees against misfuelling.

A natural successor rather than a complete reinvention, the new Mondeo is not as exciting a design as some of its teaser concepts had perhaps led people to expect. Nevertheless, it is highly accomplished.

K PR 2074

Ford Mustang

Design	Italdesign-Giugiaro
Engine	4.6 V8
Power	373 kW (500 bhp)
Gearbox	5-speed automatic
Installation	Front-engined/rear-wheel drive
Front suspension	MacPherson strut
Rear suspension	Rigid axle
Brakes front/rear	Discs/discs
Front tyres	275/40R19
Rear tyres	315/35R20
Length	4703 mm (185.2 in.)
Width	1990 mm (78.3 in.)
Height	1363 mm (53.7 in.)
Wheelbase	2720 mm (107.1 in.)
Track front/rear	1676/1676 mm (66/66 in.)
Kerb weight	1500 kg (3307 lb.)

Fabrizio Giugiaro has long been a fan of Ford's current production Mustang; moreover, more than a generation ago – in 1965, not long after the original's launch – his father, Giorgetto Giugiaro, designed a one-off special Mustang for Bertone.

This new concept is in many ways similar. It, too, is based on a current production Mustang, and it is also said to have come about because the Giugiaro family liked the standard design. Fabrizio's interpretation is smoother and sleeker, with some of the Stateside muscle making way for Italian finesse, especially in the details. The rear overhang is shorter; the body is wider, especially in the centre, where fashionable butterfly doors appear; and a more dynamic look is created by the faster windscreen, which stretches right over the roof to the back of the car, the glazed area stopping just short of a spoiler tip.

The deep clamshell bonnet wraps right over the sides to create a shut line that leads into the door recess feature. The DLO along the car winds up and down, accentuating the power at the wheels and adding a sense of energy that threads its way through the whole car, something that many argue is missing in the production Mustang. The triple rear lamps are a direct reference to the original 1960s Mustang, as is the shape of the grille.

The interior of this car is fabulously overstated, with bold fur piping and panels that contrast well with the orange-and-brown leather upholstery and the orange exterior paintwork. Outrageous, perhaps, but certainly more fun than the current macho production interior.

This partnership between Jay Mays, head of design at Ford, and Italdesign has sought to re-energize the Mustang brand. There is just one headache: they have done such a great job that Mustang dealers across America will soon be fed up explaining to eager buyers that it is only a one-off show special.

STANG

Honda Accord Coupé

Coupé derivatives do not normally feature on their own in the *Car Design Yearbook*, being merely body variants of the baseline car. The Accord Coupé concept is different, however. The Accord is Honda's bestselling model in North America, and an all-new eighth-generation model is due for release shortly after this book is published.

With an eye to gaining kudos over its rivals, Honda has seized the opportunity in this coupé concept to preview the design theme of the new Accord saloon production model. This debut at the North American International Auto Show is the first time that an Accord concept vehicle has ever been shown at a major auto show.

When Honda first launched the Accord back in 1976, it was all about comfort and practicality. Now, though, tapering lines and a long fastback roof profile that could easily have been lifted from the new Audi TT give the coupé a modern and dynamic profile. Lamps front and rear sweep around the side of the car and initiate the forward-sloping lines. The front view is constructed of regular geometric shapes, as is the rear, where the boot lid slopes back culminating with the nowadays *de rigueur* high-performance signature of quadruple exhaust tailpipes and a structure styled to resemble an aerodynamic diffuser.

There is no interior to this car as it is an exterior model only, hence the blackened windows. Honda claims it to be a five-passenger car, but judging by the fast roofline, the head room for rear-seat passengers is likely to be compromised. We can expect the production model to share this mix of dynamic lines and simple contemporary forms when it launches.

Engine	3.0 V6
Installation	Front-engined/front-wheel drive
Brakes front/rear	Discs/discs

Accord

Honda Remix

Designed at Honda's studios in California, the Remix explores the exterior themes for a new, compact two-seater coupé and is built on a front-wheel-drive platform. Superficially, it is as soft and friendly as the Acura Advance sedan concept is hard-edged and aggressive.

The Remix features a whole raft of contrasts, from the distinctive front end dominated by the large, flat clamshell bonnet and open, toothless mouth grille, to the squarer, almost estate-car-like rear end. The leading edge of the bonnet gives a semicircular feature line when viewed in plan: that feature line continues along the DLO before sweeping up in an attractive arc to meet the trailing edge of the door.

The deep grille cuts down into the body-coloured bumper, visually planting the car on the road. At the sides of the bumper, semicircular vents echo the design of the bonnet, and are copied in a slimmed-down form behind the front wheel arch and just ahead of the rear wheel arch in a rather simplistic way that undermines the originality of the overall proportion.

The streamlined windscreen is fast and aerodynamic, whereas the rear of the car – with its straight-sided tailgate glass linking the roof and the flat rear panel – looks bulky and heavy. The chunky, arched A-pillar blends with the windscreen colour to give a wraparound effect, while the elegant eye-shaped door glass would require a complex mechanism to swing it down into the door. Other notable contrasts include the compact circular rear lamps mounted far forward amid the heavy rear fenders.

Honda has tried to combine a sports look with practicality on the Remix, and has shown just how difficult it is from a design standpoint to satisfy both values. The BMW Z3M of the 1990s had a similarly heavy proportion at the rear, yet, despite its prodigious power, it received little praise when launched.

Design	Ben Davidson
Gearbox	6-speed manual
Installation	Front-engined/front-wheel drive

Concept

Honda Small Hybrid Sports

Design	Fabio Miniati, Akio Fumiiri
Engine	4 cylinder petrol/electric hybrid
Gearbox	CVT
Installation	Front-engined/front-wheel drive
Front tyres	165/60R20
Rear tyres	165/60R20
Length	4000 mm (157.5 in.)
Width	1760 mm (69.3 in.)
Height	1270 mm (50 in.)
Wheelbase	2350 mm (92.5 in.)

Though Honda exhibited the Small Hybrid Sports at the Geneva show in 2007, the model had blacked-out glass areas and no interior; uncharacteristically, the company was reluctant to say anything about the car's engineering or its future production plans. However, as an efficient hybrid-powered two-seater coupé, it could be seen as a possible blueprint for a successor to the original Insight; as such, it could grace our streets in 2009. Indeed, this new concept could be a pointer towards one of the high-volume hybrids that Honda promised in its strategy presentation of 2006.

Designed in Germany, the concept shows more imagination in its shape than its name; it is certainly both conceptual and futuristic, and if Honda is brave enough to market it by 2009, it will need to make tough decisions on many of the more ambitious aspects of the design.

This is clearly a design focusing strongly on low drag (though Honda will not reveal figures). It has tall but very narrow tyres sitting tightly in the wheel arches, the waisted body sits low to the ground and is absent of drag-inducing door mirrors, and the unique tyre design overlaps the wheel rims, blurring the distinction between tyre and wheel. The one-piece glass windscreen and roof and the LED-rich front-end design and unconventional bumper would also be questionable for volume production.

As a hybrid, the new coupé is sure to combine the most advanced petrol and electric technologies for impressive economy. From the concept's exterior shape it is hard to see how Honda will package these mechanical elements in the eventual production car, but that is perhaps part of the intrigue that will ensure that technology addicts are among the first to place their orders for what – if built – promises to be one of the most advanced small cars ever conceived.

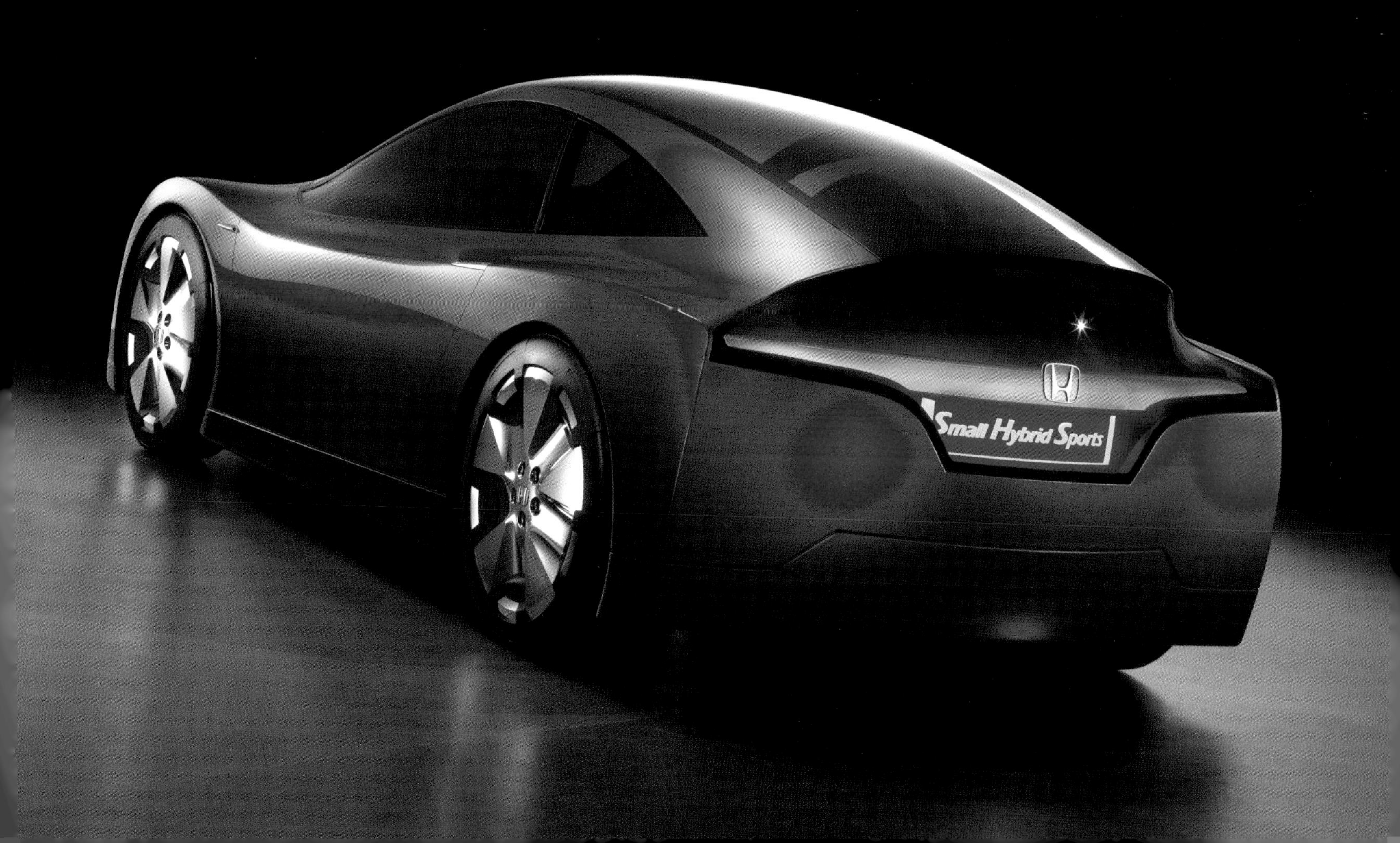
Small Hybrid Sports

Honda Step Bus

On a superficial level it might seem that this compact and boxy concept from Honda has its sights set on the equally boxy – and highly successful – Toyota Scion xB. But Honda does in fact already have a strong presence in super-practical vanlike models. The larger Element has been a big hit in the US, while in Japan the Stepwgn is still highly fashionable among young buyers, giving Honda significant credibility in this sub-niche.

Using a layout familiar from Honda's tiny micro vans, the Step Bus is around half a metre (20 in.) shorter than the Scion. The small three-cylinder engine is centrally mounted under the floor and drives the rear wheels. All four wheels are small, minimizing the encroachment on passenger and cargo space. This alone ensures that it is an effective city vehicle.

The simplicity of the Step Bus is what makes it stand out. With so many other cars on the market having complex surfacing and flowing lines, its planar, boxlike appearance clearly distances it from competitor products. It is well proportioned, and such details as the exterior lamps and the interior are all well resolved.

The Step Bus has excellent load capacity for its length, and the long sliding doors on each side make it easy to access the interior space. A raised seating position makes for good visibility, and the bluff front end makes it easier to park; this might compromise the aerodynamics, but that is hardly a top priority in a city model.

The interior has a modular construction so that it can be configured to suit the customer. Navigation systems are inbuilt, while digital music systems can be brought into the car and integrated with the built-in networks.

This is a vehicle with many different potential uses, and, like the Element before it, the Step Bus is sure to help Honda connect with the younger buyers it labels as the MP3 generation.

Design	Dave Marek
Installation	Mid-engined/rear-wheel drive

HONDA

Hyundai HCD10 Hellion

Design	Joel Piaskowski
Engine	3.0 V6 diesel
Power	176 kW (236 bhp)
Torque	450 Nm (332 lb. ft.)
Gearbox	6-speed automatic
Installation	Front-engined/all-wheel drive
Brakes front/rear	Discs/discs
Front tyres	275/55R20
Rear tyres	275/55R20
Length	4171 mm (164.2 in.)
Width	1890 mm (74.4 in.)
Height	1559 mm (61.4 in.)
Wheelbase	2555 mm (100.6 in.)
Track front/rear	1542/1542 mm (60.7/60.7 in.)

The Hyundai design team appears to have thrown the rule book out of the window in creating the HCD10 Hellion concept. Even its central notion of combining the speed and style of a sports car with the rugged, go-anywhere toughness of a 4x4 crossover is a baffling one; add to this the idea that the whole design is inspired by surfboards and hard-shell backpacks, and it becomes clear that Hyundai has opted for a unique set of design values.

Short and chunky, the Hellion resembles a muscular animal, with its bulging ribs that stretch not just horizontally but also vertically up the pillars. These give the appearance of an exoskeletal frame covered with a vacuum-wrapped skin to provide a roll cage effect. The side windows have a Gaudí-esque appearance and jar with conventional sports-car body language. Hyundai claims the Hellion is designed to have the feel of a living and breathing organism, with high off-road performance to match: this is visually suggested in many ways, from the bulging wheel arches to the stepped bonnet with its high-mounted air intake for the diesel turbo engine.

The design gets more unusual still the further rearward one looks, culminating in the oddly shaped tailgate and the rear lamps that jut out, motorcycle-style, from the base of the rear quarter-windows.

The framelike theme is carried through to the interior. Front occupants are given deep seats and the environment is both protective and sporty, with matt aluminium and black the dominant colours.

This highly individual vehicle would certainly stand out and could even generate a niche following from those looking to make a unique – if impractical – statement. But it would be a brave company indeed that decided to build it.

HCD-10

Hyundai i30

Engine	2.0 in-line 4 (1.4 and 1.6, and 1.6 and 2.0 diesel, also offered)
Power	143 kW (192 bhp) @ 6000 rpm
Torque	186 Nm (137 lb. ft.) @ 4600 rpm
Gearbox	5-speed manual
Installation	Front-engined/front-wheel drive
Front suspension	MacPherson strut
Rear suspension	Multi-link
Brakes front/rear	Discs/discs
Front tyres	185/65R15
Rear tyres	185/65R15
Length	4245 mm (167.1 in.)
Width	1775 mm (69.9 in.)
Height	1480 mm (58.3 in.)
Wheelbase	2650 mm (104.3 in.)
Track front/rear	1546/1544 mm (60.9/60.8 in.)
Kerb weight	1252 kg (2760 lb.)
Top speed	205 km/h (127 mph)

Hyundai and its subsidiary partner, Kia, are getting serious. Not only are the two brands generating an impressive output of imaginative concept models, but also the cars in the showrooms are getting better by the day. Most recently, the combine has signalled its intentions towards Europe, commissioning dedicated designs and building extensive R&D and manufacturing facilities.

Hyundai is not afraid of competition, either: the new i30, aimed straight at the best in the VW Golf class, is a production version of the concept that previewed last year as the Arnejs. According to Hyundai, the 'i' stands for inspiration, intelligence and innovation. And, starting with the i30, the 'i' is expected to prefix the names of all forthcoming small Hyundais.

Built in the Czech Republic, the i30 is designed specifically for the European C-sector and replaces the lacklustre Accent. It is based largely on the Kia Cee'd, with which it shares much of its hardware.

The production i30 is an attractive and well-proportioned hatchback with a friendly face, and features modern crisp lines that run down its body out into the rear lamps, defining it as a confident and newly designed car. At the base of the doors the feature-line crease sweeps up in a way that echoes the line created by the quarter-window. When viewed from the rear the car is more imposing than most in its class; the large red lamps verge on the overbearing. The interior is fine for a car that represents the mid-market and is neither overly sporty nor dull.

Overall, though the i30 seems to draw in themes from many contemporary European-designed cars, it does not project a strong identity of its own. With this model Hyundai is playing the long game, cautiously, just as Toyota does, in the knowledge that buyers in search of a more exciting style rather than a car that simply looks modern will in any case shop elsewhere.

HYUNDAI i30

HYUNDAI
i30
HYUNDAI i30

Hyundai Quarmaq

Design	Thomas Burkle
Engine	2.0 in-line 4 diesel
Power	127 kW (170 bhp)
Gearbox	4-speed automatic
Installation	Front-engined/four-wheel drive
Brakes front/rear	Discs/discs
Front tyres	285/35ZR22
Rear tyres	285/35ZR22
Length	4484 mm (176.5 in.)
Width	1980 mm (78 in.)
Height	1622 mm (63.9 in.)
Wheelbase	2680 mm (105.5 in.)
Track front/rear	1650/1650 mm (65/65 in.)

Like so many other recent concepts, Hyundai's Quarmaq is a crossover. What makes it different is that it goes beyond the normal aim of combining sportiness with practicality and off-road ability: developed in partnership with GE Plastics, the Quarmaq has a mission to demonstrate the future potential of a whole spectrum of advanced automotive plastic materials.

A prime example is the use of a special plastic front end to offer protection to pedestrians hit by the vehicle; in the EU, says Hyundai, one-fifth of traffic fatalities are pedestrians. The vehicle also serves to demonstrate how plastic can be formed into shapes not possible with glass or metal; this is just the kind of freedom that designers relish, and the result, though highly conceptual, shows how imaginative and progressive a design can be when released from conventional constraints.

Highly distinctive design elements on the Quarmaq include the dark C-shaped door window that separates the cream-coloured front from the champagne-shade rear bodywork; from the back, there is a third colour in the shape of the central black rear window that tucks between the back lights and broadens out to form the rear apron. This is truly a car of two halves, with only the roof and the heavy door frame bridging the intersection between the front and rear portions. In fact, the design is easier to understand with the doors open, when its shape, proportions and configuration become less confusing. Among its most dramatic features is the windscreen, with its high degree of curvature cutting straight into the door glass, which then wraps round again in a floor-level slot pointing at the front wheel centres.

Plastic materials also help to save weight and make the most of the environmental credentials of the 2-litre diesel engine. As an indicator of future trends the Quarmaq is certainly fascinating, if not easy to fathom or – to today's eyes at least – pretty.

Hyundai Veracruz

Design	Joel Piaskowski
Engine	3.8 V6
Power	194 kW (260 bhp)
Torque	349 Nm (257 lb. ft.)
Gearbox	6-speed automatic
Installation	Front-engined/all-wheel drive
Front suspension	MacPherson strut
Rear suspension	Multi-link
Brakes front/rear	Discs/discs
Front tyres	245/60R18
Rear tyres	245/60R18
Length	4840 mm (190.5 in.)
Width	1945 mm (76.6 in.)
Height	1750 mm (68.9 in.)
Wheelbase	2805 mm (110.4 in.)
Fuel consumption	11.8 l/100 km (24 mpg)

Hyundai's upmarket ambitions are pretty transparent with its latest offerings: the BMW-like Genesis concept shown at the New York show, and this, a substantial crossover SUV that takes a direct tilt at the Lexus RX and even the BMW X5 and the Mercedes ML-Class.

Christened Veracruz after the Mexican coastal state, this is by US standards a mid-size vehicle, which will provide seating for up to seven passengers; in Europe, where it is unlikely to appear, it would be counted in the large category. To avoid confusion, Hyundai has decided to invent a new sector for it, called the Luxury Utility Vehicle (LUV).

In Hyundai's strategy the 2008 Veracruz is a significant upward jump from the utilitarian, truck-based Terracan, which it replaces. The new model is built on a stretched version of the Santa Fe 4x4 platform, thus giving it much-improved ride and refinement and a more carlike appearance.

The difference between the Veracruz and the Terracan is total. The new car has flowing-form surfaces and a faster windscreen to give the effect of an aerodynamic shape; the bulging wheel arches of the Terracan have been eliminated, so the vehicle now looks much more like a town car for occasional trips on rugged terrain, rather than vice versa. The design looks well integrated, with chrome detailing to give an upmarket appearance. Inside, customers are presented with a high-end luxury cabin fitted with decent materials, sweeping, relaxing curves and harmonious colours.

This is to be Hyundai's biggest crossover SUV until the large Portico launches in 2009. There is some speculation that the Veracruz is testing the waters for a new luxury brand by Hyundai, to compete with the likes of Lexus, Cadillac, Lincoln, Infiniti and Acura; whatever the truth of these rumours, the model certainly marks a strong shift upmarket.

Infiniti G35

Engine	3.5 V6
Power	228 kW (306 bhp) @ 6800 rpm
Torque	364 Nm (268 lb. ft.) @ 5200 rpm
Gearbox	6-speed manual
Installation	Front-engined/rear-wheel drive
Front suspension	Double wishbone
Rear suspension	Multi-link
Brakes front/rear	Discs/discs
Front tyres	225/50R18
Rear tyres	245/45R18
Length	4750 mm (187 in.)
Width	1773 mm (69.8 in.)
Height	1453 mm (57.2 in.)
Wheelbase	2850 mm (112.2 in.)
Track front/rear	1519/1519 mm (59.8/59.8 in.)
Kerb weight	1602 kg (3532 lb.)
Fuel consumption	10 l/100 km (28 mpg)

Infiniti's new G35 represents an evolution of the existing model rather than an outright abandonment of what it has stood for. The platform may be enhanced and it uses the next-generation engines, but the design changes are subtle and the overall aura, though more distinctive and stylish, is familiar.

The G series is by some margin the biggest-selling product of Infiniti, Nissan's premium North American brand tasked with leading the fight against Lexus, BMW and Audi. Conscious of the need to maintain continuity and hence brand loyalty, Infiniti has taken a leaf out of the Germans' book by making only minor adjustments to the G35's overall look: clearly new are the fashionably sweeping, irregular-shaped lamps both front and rear, along with the twisted bars that make up the grille, inspired, Infiniti says, by traditional Japanese swords.

As before, the cabin has a rearward bias, the long rear door window giving an executive appearance in line with the model's market orientation. Over the rear wheels that propel the G35, tidy flared wheel arches sit below a curvaceous shoulder that plants the car on the road. The short boot lid now incorporates a subtle spoiler that finishes the tail off neatly.

The interior design is very conventional, to say the least, but the colours are soft tones for warmth and harmony rather than the Germanic black that many competitors go for. The arrangement of the dashboard controls is easy to use and understand, but there is little that is radical. There are some Japanese elements: violet accents on the instruments represent the Japanese colour for royalty, while '*washi*'-style textured aluminium features on the centre console, *washi* being a special handmade paper.

Well proportioned but neither bold nor exciting, the G35 is seen as a competent, if uncharismatic, competitor to the BMW 3 Series and the Audi A4.

Italdesign Vadho

Design	Italdesign-Giugiaro
Engine	BMW V12 hydrogen power
Gearbox	7-speed automatic
Installation	Side engined/rear-wheel drive
Front suspension	Double wishbone
Rear suspension	Double wishbone
Brakes front/rear	Discs/discs
Front tyres	275/35R20
Rear tyres	315/35R20
Length	4550 mm (179.1 in.)
Width	1980 mm (78 in.)
Height	1150 mm (45.3 in.)
Wheelbase	2800 mm (110.2 in.)
Track front/rear	1700/1660 mm (66.9/65.4 in.)
Kerb weight	1150 kg (2535 lb.)

Italdesign-Giugiaro's striking Vadho concept certainly got plenty of attention at the Geneva Motor Show, but not all of that attention was as complimentary as Giugiaro would perhaps have liked. There was something rather familiar about the offset tandem seating arrangement – in fact it was originally seen *chez* Giugiaro with the Aztec of 1988 – and the general demeanour of the vehicle appeared to bring little new to the party.

The offset, aircraft-like cockpit is not the most sociable of layouts, so Giugiaro thoughtfully provides a video system so that 'pilot' and passenger can see each other's faces on the move. The rationale for the side-mounted cockpit is that the BMW V12 engine and its hydrogen fuelling system can fit alongside, improving the balance of the vehicle's handling, if not that of its aesthetics.

Because the Vadho has a bubble polycarbonate roof, the car appears low and long. When the car is viewed from the front, the domed roof contrasts sharply with the narrow triangular headlamps, with their many stacked parallel elements. The single large gullwing door is a feat of engineering that opens up to display a snug, race-inspired interior, with multiple shades of grey and silver giving the whole car a sophisticated edge. Fittingly for an aircraft-like environment, the cockpit has no steering wheel but instead uses two drive-by-wire joysticks fixed on the armrests of the pilot's seating and pedals unit. The upholstery materials chosen for the interior are in metallic and phosphorescent opaque grey tones.

Giugiaro's rebranding of Italdesign can clearly be seen on the large G featuring on the front and on the door tread plate, but as a platform for launching the group's new identity such a celebrated master as Giugiaro could surely have come up with something better than an essentially familiar old concept warmed up with a few token modern details.

VADHO

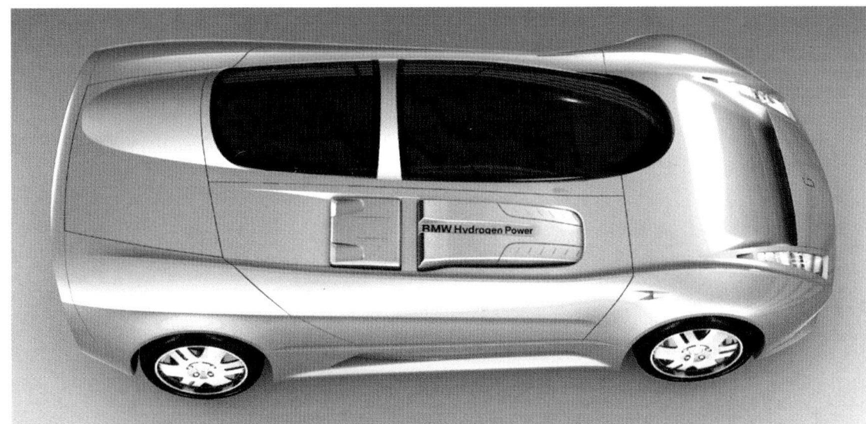
BMW Hydrogen Power

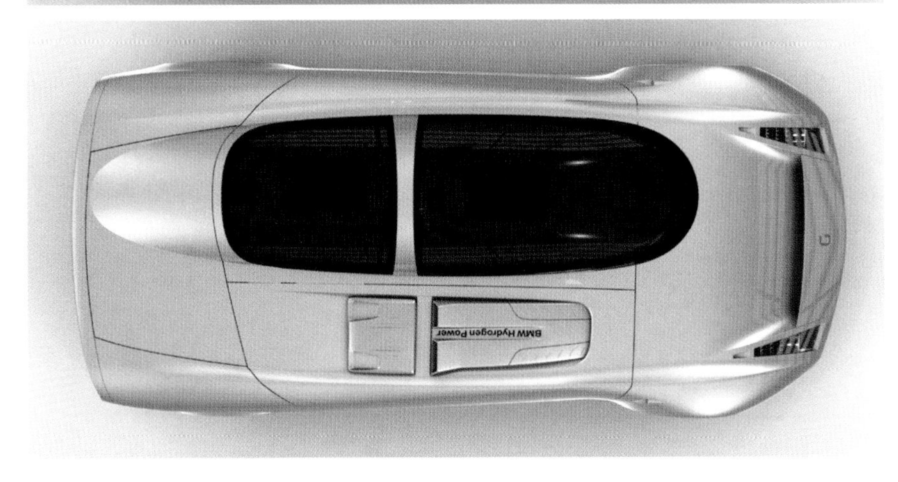
BMW Hydrogen Power

Jaguar C-XF

Design	Ian Callum
Engine	4.2 V8
Power	313 kW (420 bhp)
Torque	499 Nm (368 lb. ft.)
Gearbox	6-speed automatic
Installation	Front-engined/rear-wheel drive
Brakes front/rear	Discs/discs
Front tyres	255/30R21
Rear tyres	285/30R21
Length	4930 mm (194.1 in.)
Width	1970 mm (77.6 in.)
Height	1570 mm (61.8 in.)
Wheelbase	2905 mm (114.4 in.)
Top speed	290 km/h (180 mph)

The C-XF will come to be seen as a landmark design in Jaguar's history. The iconic British brand has been suffering a life-threatening collapse in its sales as buyers have rejected the old-fashioned image of its retro-focused product range. The C-XF is the first of a promised series of radical new Jaguars for the twenty-first century.

Ian Callum, head of Jaguar design, has already begun changing things, as can be seen by the recent and well-received XK coupé. But this new C-XF concept goes much further, seeking to stun and surprise rather than soothe the old guard hoping to see the heritage look recycled yet again. Tellingly, Callum believes that the last significant step in Jaguar's design history was in 1968, with the original XJ6. His professed goal in designing the C-XF was to imagine where Jaguar would be today if its style had continued to evolve at the rate of change shown in the 1950s and '60s.

Like those remarkable earlier designs, a contemporary Jaguar should evoke 'instant desire' by making people stop and pay attention, believes Callum; Jaguar aesthetic values continue to be purity, dynamism, latent power, balance and modernity, with the last being the most important. And the C-XF is nothing if not modern: references to Jaguar's past are minimal, but the poise of the car is still very Jaguar.

The prominent front grille is deep and gaping, and it is from this that the powerful flowing bonnet emanates, together with the aggressive-looking headlamps. The saloon-masquerading-as-coupé side profile has similarities to the Aston Martin DB7, also penned by Callum; the very rakish rear, perhaps the design's weakest point, has an excess of horizontal lines.

Innovations inside are too many to mention. As with the radical exterior, the cabin marks a complete break with Jaguar's conservative recent past and emphasizes what a bold move this design – which is a blueprint for the production XF that replaces the S-type – actually represents.

JAGUAR

Jeep Patriot

Design	Trevor Creed
Engine	2.4 in-line 4
Power	128 kW (172 bhp) @ 6000 rpm
Torque	224 Nm (165 lb. ft.) @ 4400 rpm
Gearbox	5-speed manual
Installation	Front-engined/four-wheel drive
Front suspension	MacPherson strut
Rear suspension	Multi-link
Brakes front/rear	Discs/discs
Front tyres	205/70R16
Rear tyres	205/70R16
Length	4410 mm (173.6 in.)
Width	1755 mm (69 in.)
Height	1636 mm (64.4 in.)
Wheelbase	2635 mm (103.7 in.)
Track front/rear	1520/1520 mm (59.8/59.8 in.)
Kerb weight	1410 kg (3108 lb.)
Fuel consumption	8.4 l/100 km (33.8 mpg)

Almost identical to and with the same overall dimensions as the concept version first shown six months earlier at the Frankfurt show in 2005, the production Jeep Patriot features a few minor tweaks: the new car has aluminium roof rails and bumper plates front and rear, but that is about the extent of the changes.

The Patriot is more significant than its trademark Jeep looks would have one imagine. Despite appearing to be absolutely in keeping with the Jeep tradition, the Patriot is in fact a 'Jeep lite', a model aimed at those who love authentic Jeep styling but who don't need the raw off-road credentials of other, more extreme Jeeps. Accordingly, it is built on a big-volume car platform rather than a classic Jeep chassis, and there is even a two-wheel-drive version for those who just want to potter around town.

The overall shape lends itself well to maximizing interior space; an upright taildoor, flat roof and relatively upright windscreen all make for a utilitarian interior. The proportion is boxy, intermingling rectangular forms and circles. A strong shoulder line runs the length of the car, giving a sense of protection.

Ground clearance of 20 cm (8 in.) is what one would expect from a multi-surface-capable SUV, while the squared-off wheel arches hint at even higher levels of off-road performance. There is a 'trail-rated' version that gives 2.5 cm (1 in.) more ground clearance and comes with extra body sealing and high-mounted intake and exhaust, allowing it to ford nearly 50 cm (some 20 in.) of water safely.

Through this model and its sister, the Compass, Jeep aims to reach out to the mainstream market, not just in the US but also in most world markets. Time will tell how keenly image-conscious consumers take to this diluted incarnation of a brand that prides itself on authenticity.

Jeep
Jeep
PATRIOT
GOODYEAR
WRANGLER

Jeep

Jeep

Jeep Trailhawk

Design	Ralph Gilles and Nick Vardis
Engine	3.0 V6 diesel
Power	160 kW (215 bhp) @ 4000 rpm
Torque	510 Nm (376 lb. ft.) @ 1600–2800 rpm
Gearbox	5-speed automatic
Installation	Front-engined/four-wheel drive
Front suspension	Solid axle with four-bar links
Rear suspension	Solid axle with five-bar links
Brakes front/rear	Discs/discs
Front tyres	305/45R22
Rear tyres	305/45R22
Length	4858 mm (191.3 in.)
Width	1968 mm (77.5 in.)
Height	1761 mm (69.3 in.)
Wheelbase	2946 mm (116 in.)
Track front/rear	1631/1631 mm (64.2/64.2 in.)
Kerb weight	1769 kg (3900 lb.)
0–100 km/h (62 mph)	9.0 sec
Top speed	199 km/h (124 mph)

Imagine a racy, low-rider version of a big and powerful 4x4; Ranger Rover did it a few years back with the Range Stormer concept, which led to the highly successful Range Rover Sport production model. Not to be outdone, Jeep has crossed its tough Wrangler platform with the upper half of the svelte Grand Cherokee, and added a semi-convertible top for good measure.

The outcome of the match is the Trailhawk concept, based on the long-wheelbase frame of the four-door Wrangler Unlimited. The idea, says Jeep, is to create a concept with full off-road credibility, and marry it with levels of interior comfort found on luxury SUVs. The lower part of the body has striking squared-off wheel arches with cavernous space underneath for serious off-roading. The extremely high waistline splits a distinctly angular and boxy lower body with a curved and styled upper architecture. The effect is that of a Range Rover on steroids, with the huge wheels and long bonnet making the wind-screen and rear glass look tiny.

The Trailhawk name, says Jeep, was derived from the bird-of-prey scowl given by the front-end graphics, in particular the front lamps, which peer out from under the bonnet edge. With no upper B-pillars and a removable cant rail and roof glass panels, the Trailhawk turns into an open-air T-bar (fixed central roof bar) concept. The designers note that because of the large front-axle-to-dashboard dimension, the vehicle is given an added sense of motion, with the front occupants seated back in the centre line of the car.

The interior uses a simple geometric design with a circular theme. Luxury materials are used but the design is more playful. An aluminium cross-member runs across the dashboard and a central spine runs front to back between the seats.

This concept car has great presence and should provide inspiration for manufacturers currently producing the many me-too SUVs on the market.

Jeep

Kia Carens

Engine	2.0 in-line 4 (2.0 in-line 4 diesel also offered)
Power	106 kW (142 bhp) @ 6000 rpm
Torque	189 Nm (139 lb. ft.) @ 4250 rpm
Gearbox	5-speed manual
Installation	Front-engined/front-wheel drive
Front suspension	MacPherson strut
Rear suspension	Multi-link
Brakes front/rear	Discs/discs
Length	4545 mm (178.9 in.)
Width	1820 mm (71.7 in.)
Height	1720 mm (67.7 in.)
Wheelbase	2700 mm (106.3 in.)
Track front/rear	1573/1569 mm (61.9/61.7 in.)
Kerb weight	1519 kg (3349 lb.)
0–100 km/h (62 mph)	11 sec
Top speed	190 km/h (118 mph)
Fuel consumption	7 l/100 km (33.6 mpg)
CO_2 emissions	201 g/km

The all-new Kia Carens is larger than the outgoing model, a key move that Kia's European promoters hope will allow the latest car to compete on level terms with such class-leading European and Japanese mid-size MPVs as the Vauxhall/Opel Zafira, the Volkswagen Touran, the Renault Scenic, the Toyota Verso and many others. And like many of its new-found opponents the new Carens is much more MPV than SUV and comes in a choice of a five-seater or a seven-seater version. Unlike the Scenic and the Citroën C4 Picasso, however, it has the same wheelbase and exterior dimensions in both its iterations.

The new car has a crisper, more modern silhouette, with fresh detailing and bolder touches, such as grille, lights and window graphics. Aerodynamics are better, too, with the Cd improving to 0.32 from 0.35, despite a greater frontal area. When the car is viewed from the front, the four-sided grille (reminiscent of the grille on the small and successful Picanto hatch) gives a solid, heavy look, with the roof bars and headlamps adding an upmarket touch.

Viewed from the side, the window line dips behind the A-pillar before stepping up again into the C-post; this takes visual weight off the car's flanks, where there is little ornamentation apart from chunky door handles, a lower rubbing strip and the neat kicked-up rear quarter-window. At the back there is again a sharper look; the lamps and boot lid have an interesting inward step, while the rear screen glass has squared-off corners to make it look thoroughly contemporary.

Inside there is nothing radical, just well-toned fabrics and easily accessible switchgear, with a strong circular theme running through the instruments, switches, vents and speakers. Kia is hoping to double sales from those of the previous model, capitalizing on value for money rather than the lure of a well-known brand.

KIA
LD56 HCF

Kia Cee'd

Design	Peter Schreyer
Engine	1.6 in-line 4 (1.4 and 2.0 petrol, and 1.6 and 2.0 diesel, also offered)
Power	91 kW (122 bhp) @ 6200 rpm
Torque	154 Nm (114 lb. ft.) @ 4200 rpm
Gearbox	5-speed manual
Installation	Front-engined/front-wheel drive
Front suspension	MacPherson strut
Rear suspension	Multi-link
Brakes front/rear	Discs/discs
Front tyres	205/55R16
Rear tyres	205/55R16
Length	4235 mm (166.7 in.)
Width	1790 mm (70.5 in.)
Height	1480 mm (58.3 in.)
Wheelbase	2650 mm (104.3 in.)
Kerb weight	1355 kg (2987 lb.)
0–100 km/h (62 mph)	10.8 sec
Top speed	193 km/h (120 mph)
Fuel consumption	6.4 l/100 km (44 mpg)
CO_2 emissions	152 g/km

Kia is going through something of a revolution in Europe. The Cee'd is the first car in Kia's history to be designed and built in Europe; so proud is Kia (which hails from Korea) of the new model's European roots that it proudly proclaims the name Cee'd to be made up from 'CE', which symbolizes its being made in the European Community, and 'ED', which indicates that the car is a European Design created especially with European consumers in mind.

At the Geneva show in 2006, Kia presented a concept anticipating the overall look and proportions of this production car, to compete with such European big-sellers as the Golf, the Focus and the Mégane. This is an area of the market where no lapse in design, performance or quality is tolerated, and where automakers can all too easily become unstuck.

The production Cee'd is understandably less flamboyant and progressive than the concept, but there is a clear link in the shape of the grille, for example, and in the full-length shoulder line running from the rear tip of the headlight to the front edge of the tail light. It is a tight enough package, but comes across as more cautious than the concept along the side of the car. The front-end design is deliberately homogeneous, so as not to offend, but the rear lights are just as elaborate as on the concept, even if the tapering shelf below them has been dropped.

Korean cars have frequently been criticized for weaknesses in interior design and quality; this is not the case for the European-inspired Cee'd. The dashboard is kept smooth, sober and simple, with the climate, navigation and communication displays in a neat central grouping and other switches concentrated on the steering wheel. The Slovakian-built Cee'd may not be as racy as its concept namesake, but it shows that Kia is learning the European way very fast indeed.

pro_cee'd

Kia Kue

Design	Tom Kearns
Engine	4.6 V8
Power	298 kW (400 bhp)
Torque	543 Nm (400 lb.ft.)
Gearbox	5-speed automatic
Installation	Front-engined/all-wheel drive
Brakes front/rear	Discs/discs
Front tyres	265/40R22
Rear tyres	265/40R22
Length	4724 mm (186 in.)
Width	1929 mm (75.9 in.)
Height	1600 mm (63 in.)
Wheelbase	2900 mm (114.2 in.)
Track front/rear	1640/1652 mm (64.6/65 in.)

Designed at Kia's newly opened studio in California, the Kue concept made its show debut after ex-Audi man Peter Schreyer had been in the job as Kia's design chief for just eighteen weeks. The Kue concept is based around the idea of a four-seater coupé crossover, or Crossover Utility Vehicle (CUV) as Kia calls it.

In deference to its proposed role as a sporty crossover, the Kue sits raised up off the ground and has a prominent arched glass roof to emphasize its height. This roof then sweeps out to the rear of the car, where the almost-horizontal rear window tapers to a point, which itself just dips into the vertical rear panel to provide the bare minimum of rearward vision. The curved plan at the rear of the Kue makes the design highly unusual, the shape contrasting strongly with the long, pointed door window.

Deep indents in the lower third of the doors are angled at either end to emphasize further the model's ride height. At the front, the lamps extend from the grille in a continuous expression; the small slats in the grille give the impression of filtering the air, and are replicated in the lower bumper. Further innovative features include the scissor-style pivoting doors and shell-like bucket seats. The interior projects a modern effect, with strong graphic features and a contrast between the aluminium and the two-tone grey-black trim. Touchpad and motion-sensing controls complete the futuristic agenda.

Schreyer says that there are many design elements here that give a hint for the future of Kia vehicles. This concept, with its 4.6-litre V8 power unit and all-wheel-drive system, makes it clear that the Korean manufacturer sees performance as one way of demonstrating its brand aspirations.

KIA

KTM X-Bow

Design	Kiska
Engine	2.0 in-line 4
Power	164 kW (220 bhp) @ 5900 rpm
Torque	300 Nm (221 lb. ft.) @ 2200–4000 rpm
Gearbox	6-speed manual
Installation	Mid-engined/rear-wheel drive
Front suspension	Double wishbone
Rear suspension	Double wishbone
Brakes front/rear	Discs/discs
Front tyres	205/40R17
Rear tyres	235/40R18
Length	3670 mm (144.5 in.)
Width	1870 mm (73.6 in.)
Height	1160 mm (45.7 in.)
Wheelbase	2430 mm (95.7 in.)
Track front/rear	1644/1624 mm (64.7/63.9 in.)
Kerb weight	700 kg (1543 lb.)
0–100 km/h (62 mph)	3.9 sec
Top speed	217 km/h (135 mph)

Compact, exotic and expected to go into production in late 2007, the X-Bow was one of the most exciting new concepts shown in Geneva. Born from a working relationship between the widely admired Austrian motorcycle manufacturer KTM and the design consultancy Kiska, the X-Bow is a highly specialized two-seater mid-engined roadster focused exclusively on the driving experience: like the UK's Lotus Elise, Caterham Seven and Ariel Atom, it seeks to deliver the thrills of a high-performance superbike to an audience committed to four wheels.

The whole look of the car is extreme and outrageous, yet also reassuringly functional. The black monocoque chassis tub cocoons the passengers and provides all the necessary structural elements to hold the suspension and engine in place. Double wishbones attach the wheels, which are separate visual entities; bright orange panels contrast fabulously with the black body and appear to float like aerodynamic aids at the front and rear.

Just as on a motorcycle, such elements as the exhaust pipe and the suspension are partially left on show to highlight the X-Bow's technical purity. There are no instruments, simply racing-car-style displays in the steering wheel, and only the gear lever and handbrake punctuate the transmission tunnel. Instead of a luggage compartment there is a storage box alongside the passenger footwell; this houses vehicle documents and an all-important cover, there being no roof or windscreen. Two helmets can also be stored in the passenger footwell.

Lotus has just followed up its Elise with the track-day-focused 2-Eleven. Like the Lotus, the X-Bow has very little in terms of driver aids or comforts, but these are neither expected nor wanted. In some ways a more modern, better-focused version of the Renault Sport Spyder designed in 1991, the X-Bow is sure to be highly prized by all those who value extreme driving thrills and absolute purity of design above everything else.

RACING

X-BOW

Lada C-Class

Could Lada possibly go the same way that Skoda went in the 1990s? A group of young Russian designers obviously thinks so: the Lada C-Class concept clearly shows a far sportier and more striking identity than anything we have seen previously from the Russian company.

Engineering the technology under the skin of the latest range of Lada products is a task that has been contracted out to Magna International. This should go some way to propelling Lada up towards contemporary European levels for vehicle performance and safety.

The design is very masculine, with all the sports features one could imagine. There are also some interesting contrasts: for example, the flat, undecorated bonnet is encircled by bulging wheel arches, the deep slash of the headlights on the fenders and a huge, menacing-looking grille. Big lateral air scoops in the front apron add to the racy effect.

There is some similarity in the design approach between this and the latest Scion xB. The rear end looks bulky and heavy, perhaps deliberately so in order to promote a sense of solidity in the car. The waistline rising up through the rear quarter-window adds extra focus to the passenger compartment.

Lada has long been trapped in a downward spiral of poor quality, ancient engineering and zero desirability; its wish to produce something more desirable is entirely understandable, as long as it can still also maintain its core values of affordability and fuel economy. Skoda has shown that big success can be achieved from run-down beginnings, provided the commitment, the cash and the vision are there. This concept is not a great design and it may not be precisely on target as far as vision is concerned, but it is a decent start.

Engine	2.0 in-line 4 (1.6 also offered)
Installation	Front-engined/front-wheel drive
Front tyres	225/40ZR18
Rear tyres	225/40ZR18
Length	4208 mm (165.7 in.)
Width	1837 mm (72.3 in.)
Height	1548 mm (60.9 in.)
Top speed	210 km/h (130 mph)

LADA

Lancia Delta HPE

Fiat's upscale Lancia brand has for some time been searching for a clear identity, a struggle that has seen it slip in the sales charts, especially outside Italy, and possibly face extinction.

Now, in looking to the future, Lancia is harking back twenty-seven years to its famous Delta. The neat hatchback of 1980, designed by Giugiaro, became a powerful brand icon – especially as the multiple rally-winning Integrale – and stayed in production for a decade. The new concept takes on that famous name and also parallels its genesis, borrowing the engineering elements of a volume-production Fiat, in this case the new 2007 Bravo.

The new Delta HPE, the second part of the name echoing that of another Lancia favourite, the Beta HPE high-performance estate, is a modern take on the practical sports hatchback theme. Like today's SEAT Leon and Audi A3 Sportback, it has deep sides and a generous, sloping tailgate, in the Lancia's case made largely of darkened glass.

The distinctive Lancia grille fronting the rounded nose echoes the attractive Ypsilon and Thesis, making the HPE unmistakable as a Lancia; the tail lights, drawn forward over the curve of the car's hips, are a further mark of Lancia identity. A particularly attractive feature is the chrome-rimmed side window profile that kicks up at the back, and which diverges from the side crease line as it runs rearward below the rear lamps; the roof hangs over at the rear and appears to float on the C-pillars. The concept's interior is sumptuous, with rich materials and what Lancia describes as business-class comfort worthy of a limousine, even in the rear.

Lancia promises a 100 per cent turbo engine line-up for the production Delta. More estate than coupé, this design nevertheless crosses enough boundaries to qualify as something out of the volume mainstream, and it is distinctive enough to provide the sense of identity that Lancia so badly needs.

Gearbox	6-speed manual
Installation	Front-engined/front-wheel drive
Brakes front/rear	Discs/discs
Length	4500 mm (177.2 in.)
Width	1800 mm (70.9 in.)
Height	1500 mm (59 in.)
Wheelbase	2700 mm (106.3 in.)

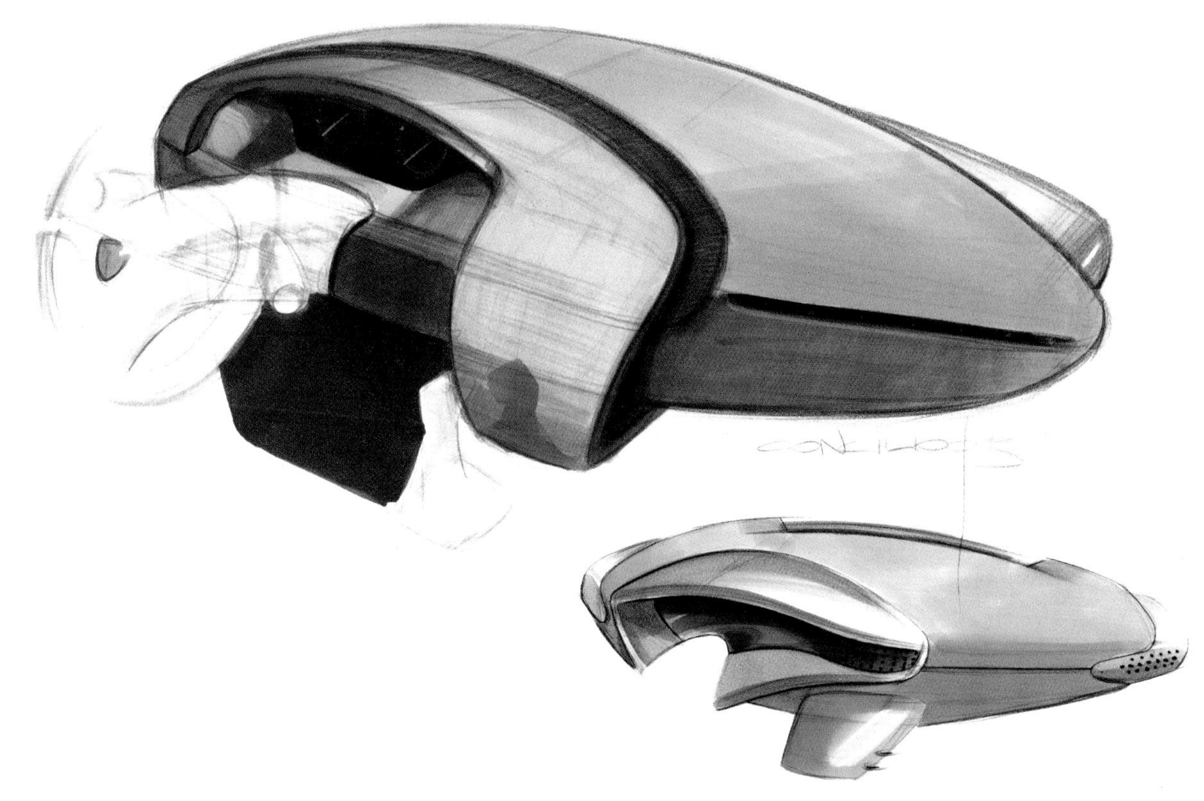

Land Rover Freelander 2

Design	Geoff Upex
Engine	3.2 in-line 6 (2.2 in-line 4 diesel also offered)
Power	171 kW (229 bhp) @ 6300 rpm
Torque	317 Nm (234 lb. ft.) @ 3200 rpm
Gearbox	6-speed automatic
Installation	Front-engined/four-wheel drive
Brakes front/rear	Discs/discs
Length	4500 mm (177.2 in.)
Width	1910 mm (75.2 in.)
Height	1740 mm (68.5 in.)
Wheelbase	2660 mm (104.7 in.)
Track front/rear	1611/1624 mm (63.4/63.9 in.)
Kerb weight	1770 kg (3902 lb.)
0–100 km/h (62 mph)	8.9 sec
Top speed	200 km/h (124 mph)
Fuel consumption	11.2 l/100 km (25.2 mpg)

As the replacement for the original that launched Land Rover as a serious 4x4 brand into the leisure SUV market in 1997, the new Freelander has a double mission. It must not only show that Land Rover has raised its game so that it is once more the clear leader in 4x4 performance, but also fit in with the new and very much classier, high-quality image being projected by today's fresh Discoverys and Range Rovers.

Fittingly, the Freelander 2 looks every bit the part. Reminiscent of a mini-Range Rover crossed with a clean-lined Discovery, it is clearly larger, classier and more sophisticated than its competitors; it looks and feels like a premium car, with the chiselled body panels being pure Range Rover and the front grille bars, echoing those on the Range Rover Sport, appearing to be crafted from solid aluminium.

The air vents in the fenders also echo those sporty models, reinforcing the upmarket impression, while the jutting forward of the grille is a clear Freelander trademark. The clamshell bonnet and large glass areas are further Land Rover identity points.

Blacked-out upper B- and D-pillars give the impression of more glass and better visibility and help visually to lengthen and lower the car; the bumpers now come body-coloured, with a lower dark-grey skirt that wraps the whole vehicle. The doors have pronounced ledges at the bottom to prevent mud and stones from being thrown up and damaging the paintwork.

Compared with the original Freelander, the interior is of much higher quality not only in terms of its architecture and materials, but also in the roominess it offers. The design language, to quote the lead interior designer, is 'strong but not intimidating'. All in all, the new Freelander is a class act that raises the bar for Land Rover's entry model and the high-quality SUV game in general.

LAND ROVER

Lincoln MKR

Design	Peter Horbury and Gordon Platto
Engine	3.5 V6
Power	309 kW (415 bhp)
Torque	543 Nm (400 lb. ft.)
Installation	Front-engined/rear-wheel drive
Front suspension	MacPherson strut
Rear suspension	Multi-link
Brakes front/rear	Discs/discs
Length	4971 mm (195.7 in.)
Width	1915 mm (75.4 in.)
Height	1339 mm (52.7 in.)
Wheelbase	2868 mm (112.9 in.)
Track front/rear	1618/1620 mm (63.7/63.8 in.)

Lincoln, Ford's premium US brand, is having a hard time establishing an identity in its struggle to compete with Lexus and a revitalized Cadillac. It has generated large numbers of concept cars in recent years; most were aggressive-looking machines springing from the pen of Gerry McGovern, and it seemed as if parent company Ford was chronically uncertain of what to do with the brand.

However, now that Peter Horbury is in charge of all Ford group design in North America, things might begin to change. The MKR concept is what Horbury tells us future Lincolns will look like: sexy and strong-looking with bold graphics and proportions that mark it out from other brands.

The MKR, based on a stretched version of Ford's Mustang platform, has certainly generated some excitement, not least because of the fact that its twin-turbo V6 engine is seen as the way ahead in terms of performance and emissions, making a production version a clearer possibility.

The concept does indeed come across as a handsome four-door coupé that might compete with the new Jaguar C-XF or the Mercedes-Benz CLS-Class of 2005. The design team looked at a whole range of memorable past Lincoln models to help provide inspiration for the MKR; yet, Horbury claims, the design is not retro in any way. Instead, it is a modern representation that gives a nod to Lincoln's history. The split, wing-shaped grille, for example, is suggestive of the 1941 Continental, while the 1961 Continental provides the smooth side profile and the rear lamps that cross the whole width of the boot. Other specific features worthy of note are the convex–concave form of the front fenders, and the dramatic interior with egg-shaped seats and long centre console that sweeps up on to the dashboard.

This is an attractive concept that suggests that Lincoln may at last have found its direction; it should be embraced and put into production with minimal change.

Lotus 2-Eleven

Design	Russell Carr
Engine	1.8 in-line 4
Power	188 kW (252 bhp) @ 8000 rpm
Torque	242 Nm (178 lb. ft.) @ 7000 rpm
Gearbox	6-speed manual
Installation	Mid-engined/rear-wheel drive
Front suspension	Double wishbone
Rear suspension	Double wishbone
Brakes front/rear	Discs/discs
Front tyres	195/50R16
Rear tyres	225/45R17
Length	3822 mm (150.5 in.)
Width	1709 mm (67.3 in.)
Height	1112 mm (43.8 in.)
Wheelbase	2300 mm (90.6 in.)
Kerb weight	670 kg (1477 lb.)
0–100 km/h (62 mph)	3.8 sec
Top speed	250 km/h (155 mph)

The Elise is by far and away the most successful car Lotus has ever made. The affordable, lightweight roadster has spawned a number of one-make race series as well as several higher-performance road-car derivatives. In essence the 2-Eleven is yet another of those derivatives, but as it is significantly different from the base car we thought it worthy of inclusion in the *Car Design Yearbook*. It could be seen as a stripped-down version of the standard Elise; we prefer to regard it as a more highly focused semi-racing car that happens to use a substantial portion of Elise expertise.

For those who remember the Renault Sport Spyder of 1992, the 2-Eleven will make perfect sense: it is a design to offer the ultimate in agility and handling thrills, for which it is essential to dispense with most of the usual creature comforts. First to go is the windscreen, helping to minimize aerodynamic drag; this limits the 2-Eleven's appeal to the truly dedicated enthusiast, as wearing a helmet is recommended.

There are two versions available, one for the road and one for the track. In the race version the driver sits in a competition-approved seat and is protected by the tubular roll bar, and the body of the car is adorned with additional aerodynamic carbon wings. The body has been stiffened with a higher-sided chassis, made possible by not having to worry about ingress and egress – there are no doors. Visually, though the front and rear lamps are from the Elise, the 2-Eleven has its own strong aerodynamic wedge-shaped profile, emphasized by the energizing graphics that run across the body.

For those who want a sure-fire track-day sensation or heightened levels of dynamic exhilaration in a compact package, the £40,000 starting cost for the roadgoing version will be but a small price to pay.

ELEVEN

Maserati GranTurismo

Design	Pininfarina
Engine	4.2 V8
Power	302 kW (405 bhp) @ 7100 rpm
Torque	460 Nm (339 lb. ft.) @ 4750 rpm
Gearbox	6-speed automatic
Installation	Front-engined/rear-wheel drive
Front suspension	Double wishbone
Rear suspension	Double wishbone
Brakes front/rear	Discs/discs
Length	4881 mm (192.2 in.)
Wheelbase	2942 mm (115.8 in.)
0–100 km/h (62 mph)	5.2 sec
Top speed	285 km/h (177 mph)

Few people would regard the Maserati GranTurismo as anything other than stunning. Designed by Pininfarina and based on the Quattroporte, Maserati's latest addition has all the characteristics of a super-stylish coupé. It replaces the Giugiaro-designed Coupé and Spyder, which were rather less voluptuous in their lines.

Maserati's first-ever roadgoing car, the A6 GranTurismo, came out in 1947, and was also designed by Pininfarina. The design of this latest car is that of a muscular beast with powerful haunches and a menacing-looking concave front grille, which will quickly consume hundreds of litres of air. Adding to the effect is the wide secondary air intake set low down below the main grille; this allows the main grille to resemble that of a 1950s racing car.

The long bonnet sits between the rising fenders and incorporates a shallow V-shaped profile running down the middle. As with the Quattroporte saloon, the three small air outlets sit behind the front wheels in their tightly fitting wheel arches. The front and rear lamp designs are new, again pointing to this car as a fresh, contemporary style full of technology but mixed with the traditional Maserati sporting character.

In the interior the leather is available in a choice of ten colours, and it is even possible to choose the colour of the carpet in the boot. Also on offer is a wide range of high-lustre interior woods that come coated with seven coats of varnish.

The GranTurismo is a welcome return to form for Maserati: it is an Italian design masterpiece of exquisite proportion, and a car big enough to carry four people in comfort and style. It is a showpiece for Italian craftsmanship, too, and to bring the point home, the options list offers a five-piece fitted luggage set designed by none other than Salvatore Ferragamo.

GranTurismo

Mazda2

Engine	1.5 in-line 4 (1.3 petrol and 1.4 diesel also offered)
Power	76 kW (102 bhp) @ 6000 rpm
Torque	136 Nm (100 lb. ft.) @ 4000 rpm
Gearbox	5-speed manual
Installation	Front-engined/front-wheel drive
Front suspension	MacPherson strut
Rear suspension	Torsion beam
Brakes front/rear	Discs/drums
Front tyres	195/45R16
Rear tyres	195/45R16
Length	3885 mm (153 in.)
Width	1695 mm (66.7 in.)
Height	1475 mm (58.1 in.)
Wheelbase	2490 mm (98 in.)
Track front/rear	1475/1465 mm (58.1/57.7 in.)
Top speed	186 km/h (115.6 mph)

At the Geneva show in 2007, the latest Mazda2 bucked the industry trend towards ever-expanding superminis: it is smaller than the model it replaces – to be precise, 40 mm (1.6 in.) shorter and 100 kg (220 lb.) lighter – and is more economical and better looking as a result.

Compared with its staid and upright Ford Fiesta-derived predecessor, the new car is far sleeker and better represents the much-hyped Mazda 'Zoom-Zoom' spirit. When viewed from the side, crisp dynamic lines are clearly visible in the rising waistline and the sculpted curve at the base of the door that leads the eye through to the rear bumper. The overall look is of a friendly and dynamic car. There is a notable simplicity of form, with such features as the grille bar, the wheels and the rear screen being strong design elements in themselves, without making the car look complex or cluttered. The headlamps and tail lamps not only bulge out but also wrap around the sides of the car, where they point towards each other along the length of the flanks, communicating through the RX-8-style crisp feature line on the crest of the front fender.

The sense of movement continues inside with contoured forms and circular features, in particular the centre heating and ventilation panel and the speedometer directly ahead of the driver. Specific interior features include the gear lever being raised to a position on the dashboard, freeing up space below for a storage compartment in the console that also incorporates a rack for maps.

Mazda's chassis engineers claim to have focused on making the car highly manoeuvrable on urban roads while maintaining high stability on motorways; it is an impressive design that not only appears well attuned to European tastes but also could stand a chance of success in a newly CO_2-conscious United States if Mazda decides to extend its reach still further.

mazda

Mazda CX-9

Engine	3.5 V6
Power	186 kW (250 bhp)
Torque	326 Nm (240 lb. ft.)
Gearbox	6-speed automatic
Installation	Front-engined/front- or all-wheel drive
Front suspension	MacPherson strut
Rear suspension	Multi-link
Brakes front/rear	Discs/discs
Front tyres	245/50R20
Rear tyres	245/50R20
Length	5070 mm (199.6 in.)
Width	1936 mm (76.2 in.)
Height	1735 mm (68.3 in.)
Wheelbase	2875 mm (113.2 in.)
Track front/rear	1654/1643 mm (65.1/64.7 in.)

From a design perspective, Mazda's new CX-9 is simply a scaled-up version of the five-seater CX-7 launched to some acclaim in 2005. However, appearances are deceptive: the CX-9 is built on a somewhat different platform, a stretched version of the CD3 architecture also used in the Ford Edge and the Lincoln MKX.

Sharing its style with the CX-7 is no handicap for the larger car. In line with their challenging brief to produce a multi-seater sports car, Mazda's designers have managed to come up with a curvaceous body for what is typically a boxy vehicle type. This does indeed give the car a sporting spirit and a softness lacking in the more macho SUVs.

Distinctive characteristics of the CX-9 include its steeply raked windscreen, rounded front end, flared wheel arches and such details as the DLO that steps up as it travels through the rear door. From the rear the CX-9 is again very attractive, with its smoothly curved rear screen and lamps that sandwich the chrome boot strip to create a full-width band.

The interior features some sporting elements: circular chrome-rimmed dials with cool-blue illumination, quality wood trim and aluminium beams with two-tone seat trim. Three trim levels are offered: Sport, Touring and Grand Touring. Standard features on all include air-conditioning, power windows, power door locks with remote keyless entry, cruise control and six airbags.

The CX-9 will be built exclusively for North America, alongside the smaller Mazda CX-7; it replaces the aged Mazda MPV. The soft, friendly and un-SUV-like shape of this design could prove a clever move at a time when the fuel thirst and imposing appearance of giant 4x4s are giving them a bad press, even in the United States.

ERM 055
CX-9

LOAD EJECT
NAVI
MENU
RTN
DISP
AUDIO
SCAN
POWER VOL
TUNE SEEK TRACK
FM/AM
SAT
CD
MEDIA
GPS
LOS ANGELES
POI
Memory Pt.
A/C
MODE
AUTO
OFF
DUAL
OUTSIDE
PASS
REAR CONTROL

Mazda Hakaze

Design	Peter Birtwhistle
Engine	2.3 in-line 4
Gearbox	6-speed automatic
Installation	Front-engined/all-wheel drive
Front suspension	MacPherson strut
Rear suspension	Multi-link
Brakes front/rear	Discs/discs
Front tyres	255/50R20
Rear tyres	255/50R20
Length	4420 mm (174 in.)
Width	1890 mm (74.4 in.)
Height	1560 mm (61.4 in.)
Wheelbase	2650 mm (104.3 in.)

The Hakaze is the third in Mazda's series of concepts to be designed around the theme of 'flow'; like the Nagare and Ryuga concepts, the Hakaze's lines are intended to evoke the flow of air or energy over its body. The word *hakaze* signifies both 'leaf' and 'air', and leaflike forms and textures can indeed be seen all over the vehicle.

Proportionally and conceptually, as a high-riding crossover coupé with an airy feel, the car is similar to the Renault Avantime: viewed side-on, it has a striking silhouette, the aluminium cant rail appearing to hang in space as the glazed upper A and B-pillars blend visually with the side glass and screen. The rear portion of the glass roof can be removed and, as on the Avantime, there is no B-pillar; the car's four frameless windows can be lowered to transform it into an open four-seat coupé.

Many influences helped to shape the Hakaze: outdoor sports, natural elements and, most clearly visible, the way in which the wind shapes sand. This can be seen in the beautifully gentle intersecting feature lines that distinguish the sides. The front-end design is very bold, with the striking aluminium plate carrying flow lines out over the headlamps and on to the fenders, giving the car the expression of a leering grin. This is echoed at the rear in the leaflike shape of the high-set rear window, the lower portion of which has leaflike ribs fanning out from the pointed ends. This point then leads to the gently curving side feature line carried forward towards the front wheel; where this intersects the feature line from the front wheel arch the surfacing is gently rippled to simulate the veins on a leaf or ripples on water.

The word in the design business is that the Hakaze's brief included the stipulation that it should be production-feasible; even stripped of some of its more outlandish features, this would be a pleasing vehicle indeed.

Mazda Nagare

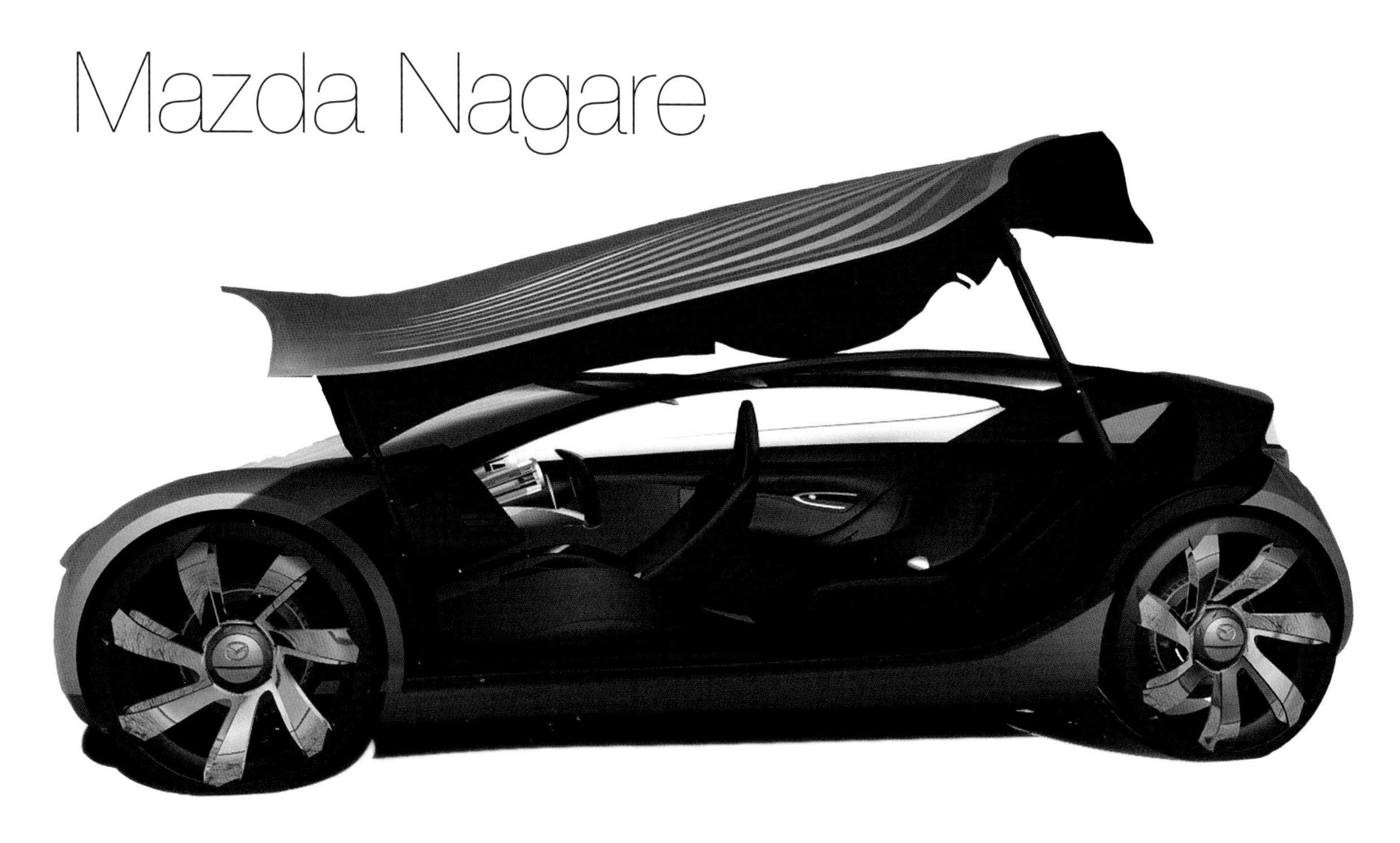

Mazda's Nagare concept is more than just the car you see; it could be described as a design philosophy statement. In fact, say its creators – Mazda's California studio – it is a concept for a concept. It is the culmination of much exploration of how to treat the surface language for a future range of vehicles, and it is this language that will inspire future concepts more directly connected with eventual production cars.

The word *nagare* is Japanese for 'flow', the embodiment of motion. Mazda's designers have striven to create and they have succeeded in doing so – a way of registering movement in their cars whether they are moving or standing still. This, the company believes, creates an energy that sits well with its 'Zoom-Zoom' mantra.

The proportion of the Nagare is sleek and streamlined: huge wheels are positioned right at the corners, while the glass canopy is inset from the doors, looking from above like a stretched oval. The nose and tail both come to a pronounced point, more rounded at the rear. A unique feature along the side of the Nagare is a series of pressed flow lines that run over the rear wheel arch; these are intended to depict the airflow along the sides of the vehicle and really add to the sense of natural movement when the car is at rest. The headlamps and rear lamps also hint at how the airflow may pass over their surfaces and fan out. Inside, the driver sits centrally, with space for three passengers on a curved sofa behind.

Incorporating this new design language into future vehicles will be a steady process. Mazda sees this as an exploration, looking towards where its designs will be in 2020. The Nagare has the purity of a design-led concept car, with engineering feasibility not really on the radar; as such, it is brilliant in the way it stretches design boundaries.

Design	Franz von Holzhausen

000

Mazda Ryuga

Design	Yasushi Nakamuta
Engine	2.5 litres
Gearbox	6-speed automatic
Installation	Front-engined/front-wheel drive
Brakes front/rear	Discs/discs
Front tyres	245/35R21
Rear tyres	245/35R21
Length	1280 mm (50.4 in.)
Width	1900 mm (74.8 in.)
Height	1260 mm (49.6 in.)
Wheelbase	2800 mm (110.2 in.)

Designed in Hiroshima by Yasushi Nakamuta, the Ryuga concept is a further Mazda design study based around the idea of *nagare*, which means 'flow' in Japanese. The earlier Nagare concept car from the Los Angeles show is also featured in this edition of the *Car Design Yearbook*, but was just an exterior model; the Ryuga, on the other hand, has a fully worked interior and develops further the idea of the 'flow' design language.

This compact four-door sports coupé shows off design ideas for a future generation of Mazda vehicles. Mazda's design chief, Laurens van den Acker, described the Ryuga as 'vibrant, confident and fun at heart. It's athletic and sporty, emotional and exotic.' The giant, full-length gullwing doors are certainly exotic, though could be discounted on cost grounds for any future volume production version.

A striking aspect of the Ryuga's side design is the set of sweeping lines on the doors, a feature inspired by the raked pebbles in a Japanese dry garden. The wheels, as well, feature spokes that twist near the hub, as if the torque of the car's imaginary flex-fuel engine has twisted them. The pearlescent paintwork, which varies from red to blue to yellow depending on the light, is meant to remind onlookers of flowing lava. Other body graphics are intended to evoke the morning dew on a bamboo leaf; the headlamps are in fact shaped like bamboo leaves.

The interior design complements the exterior very well. Exotic forms sweep through the cockpit, making this car look futuristic and very driver-focused. The partly open steering wheel and bright-blue gauges that nestle around the steering column sit within a V-shaped dashboard recess that swirls around to plunge through a slot in the centre tunnel.

All in all, this is a highly imaginative concept that points to an exciting future at Mazda.

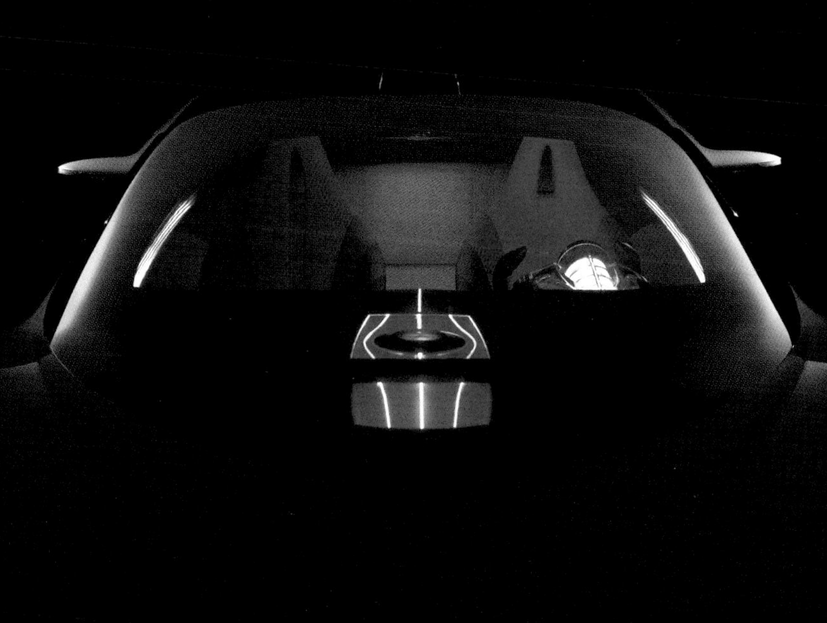

Mazda Tribute

Engine	3.0 V6 (2.3 in-line 4, and HEV, also offered)
Power	149 kW (200 bhp) @ 6000 rpm
Torque	262 Nm (193 lb. ft.) @ 4850 rpm
Gearbox	4-speed automatic
Installation	Front-engined/four-wheel drive
Front suspension	MacPherson strut
Rear suspension	Multi-link
Brakes front/rear	Discs/discs
Front tyres	235/70R16
Rear tyres	235/70R16
Length	4429 mm (174.4 in.)
Width	1828 mm (72 in.)
Height	1720 mm (67.7 in.)
Wheelbase	2620 mm (103.2 in.)
Track front/rear	1557/1548 mm (61.3/60.9 in.)
Kerb weight	1583 kg (3490 lb.)
Fuel consumption	11.2 l/100 km (25 mpg)

The 2008 model of the Mazda Tribute has had a significant revamp, but is essentially a rebadged version of the US Ford Escape. The Ford Escape has been offered with a hybrid version for almost two years; now Mazda, which is one-third owned by Ford, is able to have its first model that can benefit from this technology.

The Mazda Tribute Hybrid-Electric Vehicle (HEV) is a full hybrid, meaning that it can run on 100 per cent electric power up to 40 km/h (25 mph) for a useful distance, helping to minimize intrusion in sensitive urban areas, while the regenerative braking captures kinetic energy and stores it in the batteries for later use. This gives maximum fuel economy and ensures that the Tribute meets California's Advanced Technology Partial Zero Emissions Vehicle (AT-PZEV) requirements by achieving Super Ultra Low Emissions Vehicle II (SULEVII) standards, the strictest petrol-fuelled emissions regulations in North America.

In design terms, the Tribute's style is much less advanced than its engineering. The proportion does not have the modern look of the rest of the cars in the Mazda line-up, betraying the model's outside origins. Changes from the older model include a new grille, headlamps and tailgate, while the sides have been given cleaner lines and rounder wheel arches, again similar to the Ford Escape. The command seating position gives good visibility and the boxy shape at the rear means that the Tribute is a practical load-carrier.

While the Tribute may be a useful model for Mazda in North America, where it is essential to be present at all levels of the SUV market, it does illustrate the risks of badge engineering for commercial convenience. Saab found this out to its cost with the disastrous Subaru-built 9-2, and Mazda customers may well be excused for wondering how a boxy, undynamic American 4x4 fits within its range of shapely and well-structured Japanese-built models.

Mercedes-Benz C-Class

Engine	3.5 V6 (1.8 in-line 4, 2.5 V6 and 3.0 V6, and 2.1 in-line 4 and 3.0 V6 diesel, also offered)
Power	200 kW (268 bhp) @ 6000 rpm
Torque	350 Nm (258 lb. ft.) @ 2400–5000 rpm
Gearbox	7-speed automatic
Installation	Front-engined/front wheel drive
Front suspension	Three-link
Rear suspension	Multi-link
Brakes front/rear	Discs/discs
Front tyres	225/45R17
Rear tyres	225/45R17
Length	4581 mm (180.4 in.)
Width	1770 mm (69.7 in.)
Height	1448 mm (57 in.)
Wheelbase	2760 mm (108.7 in.)
Track front/rear	1533/1536 mm (60.3/60.5 in.)
Kerb weight	1610 kg (3549 lb.)
0–100 km/h (62 mph)	6.4 sec
Top speed	250 km/h (155 mph)
Fuel consumption	9.7 l/100 km (29 mpg)

With much of the DaimlerChrysler management distracted by the troubles of the group's under-performing Chrysler arm, the core Mercedes brand has been suffering from problems, in particular electrical issues, that have let it slip down in the quality and customer satisfaction rankings. The new C-Class is tasked with changing that once and for all: it needs to be perfect straight out of the box in order to restore confidence that Mercedes-Benz is back on top as the quality leader in the premium segment.

The C-Class is Mercedes' biggest-selling line, which raises the stakes still further. Accordingly, the vehicle development process included driving some 200 prototypes an unprecedented 24 million kilometres (15 million miles) to iron out any possible gremlins.

Like its forebear, the new C-Class looks like a scaled-down S-Class, especially in the forward slope of the tail and in the way the side surfacing is handled. For Mercedes officials, however, the big breakthrough is the decision to produce two visually different identities. The Elegance model, aimed at retaining the traditional older buyer, has the familiar Mercedes grille and bonnet-top star; the Classic and the Avantgarde, on the other hand, are targeted at winning a more youthful, BMW-buying audience and use a CLK-style coupé front with the large logo set within the three-bar grille. The interiors are differentiated in tone and texture rather than structure or fittings; like the exterior, they are evolutionary rather than innovative.

In practice the exterior differences between the versions are little more than cosmetic, as both share identical sheet metal. The Elegance generally has more chrome and glitz, the Classic and Avantgarde a lower, sportier stance; nevertheless, each represents only a slight variation on a familiar theme, and it would be wrong to imagine any deep-down differences in personality or image.

S ON 3942

Mercedes-Benz Ocean Drive

Mercedes unveiled this huge four-door S-Class-based Ocean Drive convertible at the Detroit show in January 2007. Named after the famous road on Miami Beach's South Beach, Florida, the Ocean Drive is designed to appeal to the style-conscious rich American. It comes at the same time as Rolls-Royce presents its ultra-luxurious Phantom Drophead four-seater convertible and just a matter of months after Bentley launched its second big open car, the Continental GTC (these models are not significantly design-differentiated from the original cars and are therefore not included in this book).

Based as it is on the large, long platform of the S-Class, the Ocean Drive concept is rare in combining a four-door layout with a full-length fabric convertible top (a folding hard-top of this great length was beyond the ingenuity of even Mercedes' engineers).

Compared with the standard S-Class, the front grille is larger, the LED headlamps are taken up into the tops of the fenders, and flush-mounted door handles are set into the fine shoulder line that runs the length of the car. The two shades of the subtle champagne-gold paintwork are separated by this line, highlighting the interplay between the convex and the newly introduced concave surfaces. Particularly eye-catching are the elegant 21-inch thirty-six-spoke wheels, which help to disguise the car's mass.

When the roof and windows are down, a neat tonneau panel (cover) finished in bird's-eye maple covers the stowed hood, the same finish being carried round the interior door panels to the central strip of the dashboard. The unusually sleek windscreen protects the occupants from the worst effects of the airflow, and just to ensure that the car's pampered occupants stay warm and comfortable even when travelling al fresco in the depths of winter, Mercedes' ingenious 'airscarf' system circulates heated air at head and neck level.

With two production models and this sophisticated concept all vying for the attention of the super-rich open-air élite, the money-no-object convertible appears to be enjoying an unexpected renaissance.

Engine	5.5 V12
Power	380 kW (510 bhp)
Torque	830 Nm (612 lb. ft.) @ 1900–3500 rpm
Installation	Front-engined/rear-wheel drive
Front suspension	Four-link
Rear suspension	Multi-link
Brakes front/rear	Discs/discs
Front tyres	275/35ZR21
Rear tyres	275/35ZR21
Length	5293 mm (208.4 in.)
Width	1911 mm (75.2 in.)
Height	1497 mm (58.9 in.)
Wheelbase	3165 mm (124.6 in.)
Track front/rear	1622/1628 mm (63.9/64.1 in.)

Design	Gerd Hildebrand
Engine	1.6 in-line 4
Power	130 kW (175 bhp) @ 5500 rpm
Torque	260 Nm (192 lb. ft.) @ 1700–4500 rpm
Gearbox	6-speed manual
Installation	Front-engined/front-wheel drive
Front suspension	MacPherson struts
Rear suspension	Longitudinal struts with Z axle
Brakes front/rear	Discs/discs
Front tyres	195/55R16
Rear tyres	195/55R16
Length	3714 mm (146.2 in.)
Width	1683 mm (66.3 in.)
Height	1407 mm (55.4 in.)
Wheelbase	2467 mm (97.1 in.)
Track front/rear	1453/1461 mm (57.2/57.5 in.)
Kerb weight	1130 kg (2491 lb.)
0–100 km/h (62 mph)	7.1 sec
Top speed	225 km/h (140 mph)
Fuel consumption	6.9 l/100 km (40.9 mpg)
CO_2 emissions	164 g/km

It is difficult to know how to regard this new Mini. Almost every part you see from the outside has been changed, and much of the interior is heavily revised, too; this qualifies it as a major overhaul. Yet the overall impression is exactly the same as that of the outgoing car, and you have to look very closely indeed to spot all the changes. So to many it will still be the same Mini that has become such a huge hit over the past five years.

BMW's reluctance to make changes is entirely understandable. The astonishing success of the 2001 model took even BMW by surprise; both that model and the update of 2007 owe their inspiration to the spirit of the 1959 original, so it could be argued that any departures from the style would show disrespect for the first model.

Accordingly, the few changes that are discernible are practical rather than aesthetic in nature. The most visible difference is the longer bonnet, dictated by the new BMW engine and the latest levels of pedestrian safety legislation; other changes are a consequence of this. The car is longer by 60 mm (2.4 in.), but this is disguised by more rounded contours; the headlamps are larger and are now fixed to the understructure rather than being part of the bonnet. Noticeable on the high-performance version, the Cooper S, is the air scoop in the bonnet.

Inside, the feeling is more luxurious than it was previously, but is still avant garde: the central speedometer has grown even more exaggerated in size and now carries the CD controls, while a novel circular electronic key slips into a slot behind the steering wheel to enable the start/stop button that then fires the car into action.

This is a new car that stays totally faithful to the original. The buyers are sure to like it, even if the design community is itching to see Mini progress to more radical things.

NU56 ZVW

SR56 LNE

Mitsubishi Lancer

Engine	2.0 in-line 4
Power	113 kW (152 bhp)
Gearbox	CVT with 6-speed sportronic mode
Installation	Front-engined/front-wheel drive
Brakes front/rear	Discs/discs
Length	4570 mm (180 in.)
Width	1760 mm (69.3 in.)
Height	1490 mm (58.7 in.)
Wheelbase	2635 mm (103.7 in.)

The new Lancer is the production car that evolved from the Mitsubishi Concept-X first shown at the Tokyo Motor Show in 2005. This, and the companion hatchback-format Concept Sportback, were reviewed in the *Car Design Yearbook 5*; while the Sportback concept was a pointer to the European hatchback model to appear in late 2006, this Lancer is the definitive four-door sedan version and has been on sale in North America from the spring of 2007.

Also shown at the Detroit show, to add to the confusion, was the Mitsubishi Prototype-X. An enhanced version of the four-door Lancer, this model will become the new Mitsubishi Evo, eagerly awaited by sports enthusiasts as the car that will re-enter top-class rallying.

The contrast with the outgoing Lancer is total. The new model is much more modern in profile, and has a longer, more spacious cabin and an aggressive front-end design with a sharklike nose and piercing headlamps that appear to glower from under the rim of the bonnet. Its powerful looks are emphasized further by the lower body trim that visually connects it to the ground, while the conventional four doors and reasonable ground clearance ensure that its usefulness as a practical car remains. The rear spoiler is quite prominent for an upmarket production saloon; this in itself will make the Evo version appeal to a different clientele from the typical Audi or Volvo buyers, who prefer the sporting attributes of their car to be more subtle.

Differences between the future Evo and the base car include a deeper and more aggressive front bumper, a full-depth grille, a larger rear spoiler, bigger wheels, red Brembo-branded brake calipers and an air intake in the bonnet.

Mitsubishi has a strong reputation for performance and quality. The dull design of the original Lancer has always been at odds with these values; the new car will surely make amends.

Nissan Altima

Engine	2.5 in-line 4 plus 40 hp electric motor assist
Power	148 kW (198 bhp total) @ 5200–6000 rpm
Torque	220 Nm (162 lb. ft.) @ 2800–4800 rpm
Gearbox	CVT
Installation	Front-engined/front-wheel drive
Front suspension	MacPherson strut
Rear suspension	Multi-link
Brakes front/rear	Discs/discs
Front tyres	215/60R16
Rear tyres	215/60R16
Length	4806 mm (189.2 in.)
Width	1768 mm (69.6 in.)
Height	1476 mm (58.1 in.)
Wheelbase	2776 mm (109.3 in.)
Track front/rear	1542/1534 mm (60.7/60.4 in.)
Kerb weight	1564 kg (3448 lb.)
Fuel consumption	6.2 l/100 km (45.5 mpg)

The fourth-generation Altima was launched at the New York Auto Show in 2006. This is the first vehicle to use Nissan's smaller D platform, and comes complete with new upgraded front and rear suspension systems. The size of the car is almost identical to that of the model it replaces, with a wheelbase just 2.5 cm (1 in.) shorter but no reduction in interior space. This is Nissan's leading model in the key mid-size market in North America.

From the side the profile comes across as smooth and aerodynamic, sloping gradually towards the rear, which is itself tipped with a small boot spoiler. The vehicle has a well-proportioned sporting stance, even though in its overall demeanour it does little to stand out from the crowd. An angular shoulder running through the rear three-quarter areas leads directly into the large multicoloured rear lamps. The playful nature of these lamps – which Nissan describes as 'jet-inspired' – is a surprise on what is otherwise a rather sober and conventional saloon. The headlamps are similar to those on the outgoing model; they sandwich a wider but thinner grille, the intention of which is clearly to widen the car.

Inside, the dashboard and instruments have a sporty circular theme, with a small steering wheel complete with multifunctional control buttons. The interior has a general feel of warmth and intimacy.

More interestingly, Nissan has agreed with global hybrid leader Toyota to make use of some of its existing technology in the new Altima. Designed to comply with emissions regulations in eight US states, including California, the hybrid version of the Altima will be Nissan's first-ever hybrid car. Fuel consumption is expected to be 6.2 l/100 km (45.5 mpg) in the city, and Nissan claims that 1120 kilometres (700 miles) will be possible on a tank of fuel, with regenerative braking helping to top up the batteries.

ALTIMA
NISSAN
HYBRID

SOURCE
ENTER
CANCEL
ON/OFF

Nissan Bevel

Design	Bruce Campbell
Engine	2.5 V6 hydrogen/electric hybrid
Gearbox	CVT
Installation	Front-engined/front-wheel drive
Brakes front/rear	Discs/discs
Front tyres	245/55R20
Rear tyres	245/55R20
Length	4399 mm (173.2 in.)
Width	1905 mm (75 in.)
Height	1621 mm (63.8 in.)
Wheelbase	2931 mm (115.4 in.)

Nissan's Bevel concept is startling as well as intriguing. Half van, half sports SUV, aimed at middle-aged or retired men, it looks vaguely like a military assault craft and, in truth, is just about as ugly. Despite the generally held conceit, propagated by the ancient Greeks, that symmetry is beautiful, Nissan appears to have gone out of its way to create an asymmetrical design.

The Bevel was presented at the Detroit show as a 'single-purpose, multifunction' vehicle, targeting men who can see it as both a tool and a toolbox. Tellingly, Nissan predicted that most journeys would be with only the driver on board. But the owner might find himself spending more time in the rear of the van than at the wheel, for it is set out as a hobby and DIY centre. All the seats fold flat to make a large, useful loading area, the tailboard doubles up as a workbench – careful with that circular saw! – and there are electric sockets for other power tools. The top half of the open tailgate protects the hobbyist from the rain, and the roof features small lashing hoops for tying down bulky cargo. There is even a removable dog kennel to give a pet a secure place to rest.

The Bevel's exterior is determinedly asymmetrical: the grille, bonnet and roof-mounted solar panel are all offset in some way, and the windscreen extends further into the roof on the passenger side. In terms of design language, a kind of hexagonal theme recurs throughout, echoing the hexagonal drive on socket tools.

People may scoff at its looks, but in its role the Bevel explores an interesting direction as perhaps the only vehicle aimed at a retiree wanting to build his own cabin in the woods. But even empty-nesters have aesthetic standards, and this concept is too strange by far.

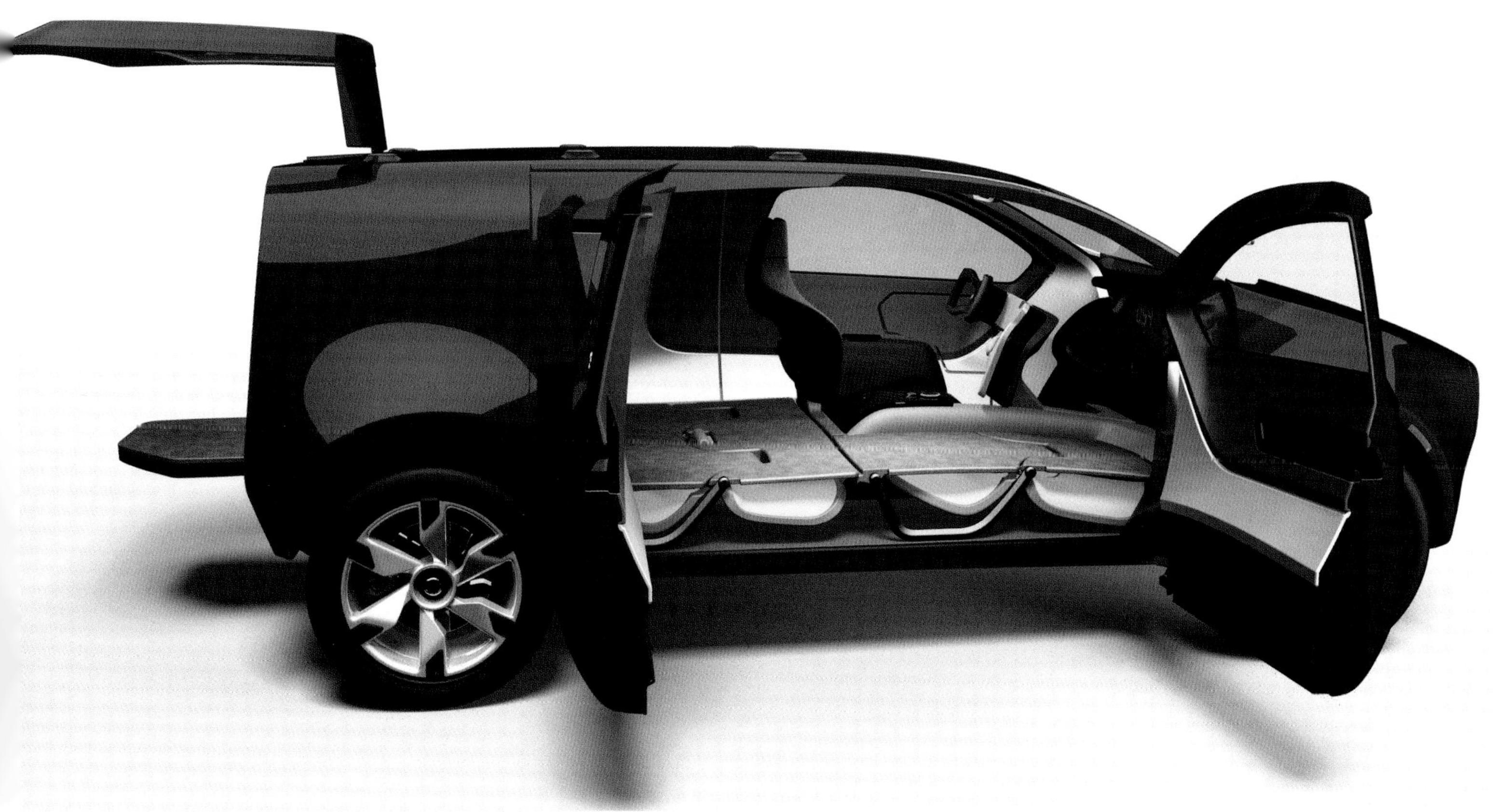

Nissan Livina Geniss

Torque	150 Nm (110 lb. ft.)
Gearbox	5-speed manual
Length	4178 mm (164.5 in.)
Width	1690 mm (66.5 in.)
Height	1565 mm (61.6 in.)
Wheelbase	2600 mm (102.4 in.)
Fuel consumption	6.3 l/100 km (44.8 mpg)

The first Nissan model to have its world debut in China, the clumsily named Livina Geniss is also the first of a new family of Nissan global cars based on a shared Nissan–Renault platform. Further minivans related to the Geniss are likely soon to appear in different markets with different exterior design themes.

A seven-seater, the Geniss – or Jun Yi, in Chinese – is built at the Huadu plant in China's Guangdong province. A relatively conservatively styled people-carrier, it has most of the practicalities of an MPV but with proportions more akin to those of a crossover, with a slightly lower roof and a faster, longer shape.

At the front the curved grille bars extend right out to the large triangular headlamps, echoing the look of the current Nissan line-up. This design uses modern surfacing to add interest to the shape: for example, the door panels are predominantly vertical in section, but a crease line towards the top creates a longitudinal shelf that helps to highlight the window area and waistline. Blackened upper door pillars are a contemporary touch, as is the large, frameless, curved tailgate glass with the currently fashionable dip in the centre. Overall, the design is very restrained, with taut surfaces that communicate a conservative emotion with little ornamentation.

Both the exterior and the spacious interior of the Geniss were designed at Nissan's technical centre in Atsugi, Japan. Although the car is locally built and China is a fast-growing market, conditions are not always straightforward and consumer preferences can sometimes lead to highly localized design features. The Geniss is in many ways symbolic of international automakers' desire to offer products better matched to Chinese tastes; whether it will succeed in China may depend more on its layout, format and pricing strategy than on the style of the sheet metal that the Japanese design studio has provided.

LIVINA
GENISS

GENISS
LIVINA
GENISS

Nissan Qashqai

Design	Stéphane Schwarz
Engine	2.0 in-line 4 diesel (1.5 diesel, and 1.6 and 2.0 petrol, also offered)
Power	110 kW (148 bhp) @ 4000 rpm
Torque	320 Nm (236 lb. ft.) @ 2000 rpm
Installation	Front-engined/front-wheel drive
Front suspension	MacPherson strut
Rear suspension	Multi-link
Brakes front/rear	Discs/discs
Length	4310 mm (169.7 in.)
Width	1780 mm (70 in.)
Height	1610 mm (63.4 in.)
Wheelbase	2630 mm (103.5 in.)

When Nissan first showed the stylish Qashqai concept at Geneva in 2004 it drew near-universal praise for its bold new proportions and elegant look. Few observers believed that the model would eventually make volume production; fewer still thought that it would reach the showrooms with so much of its style and originality intact.

But this is precisely what has happened, earning Nissan a commendation for its courage in seeing through the Qashqai's mission of attacking the volume market from an alternative tack. For it is Nissan's belief that the best route to decent profits in the Golf class is through being original and different, approaching customers not as a me-too hatchback but as a left-field dissident with unique appeal. So although the Qashqai has the allure of an adventurous traveller – the nomadic tribe from which its name comes – it is in fact standard family fare underneath, and its strategic role is to replace the humdrum Almera hatchback.

The design projects the strength of a 4x4, with its high build, pronounced wheel arches and chunky sections in the A-pillar and front grille; the concept's dramatically arched DLO is still discernible, though in less exaggerated form. Some of the sharpness of the concept has given way to rounder forms on the commercial car, and the novel vertical tail lights extending into the roof have been dropped in favour of horizontal shapes.

Even so, the Qashqai has a different proportion and a different stance from anything else in its class; equipped with a very stylish and modern interior, it promises to do the job of a hatchback as well as any, and to stir in a sense of adventure and occasion too.

One of the most progressive new production cars featured in this edition of *Car Design Yearbook*, this is a model whose progress will be tracked with interest by car companies the world over.

NISSAN
QASHQAI

QASHQAI

Nissan Rogue

Engine	2.5 in-line 4
Power	127 kW (170 bhp)
Torque	237 Nm (175 lb. ft.)
Gearbox	CVT
Installation	Front-engined/front-wheel drive
Front suspension	MacPherson strut
Rear suspension	Multi-link
Brakes front/rear	Discs/discs
Front tyres	215/70R16
Rear tyres	215/70R16
Length	4645 mm (182.9 in.)
Width	1800 mm (70.9 in.)
Height	1658 mm (65.3 in.)
Wheelbase	2690 mm (105.9 in.)

Nissan's naming of the Rogue is a cry for attention in a market crowded with SUVs that are basically very similar. The rebel Rogue is trying desperately to stand out from the crowd and assert, as Nissan terms it, 'its independence-oriented positioning'.

Nissan already has four other SUVs in its North American line-up: the Armada, the Pathfinder, the Xterra and the Murano. But such is the momentum of the swing towards crossovers that the company believes there is room for another to compete against the similarly dimensioned Hyundai Santa Fe and the Honda CR-V, in what in the US counts as the entry level of the segment. Built on the same underpinnings as the new Nissan Sentra, itself structured around the global Renault–Nissan C platform, the Rogue is in effect first cousin to the European Qashqai, also derived from the same components.

As befits a US model, the Rogue is longer and larger than its European cousin; it comes standard in front-wheel-drive layout, but four-wheel drive is an option. The proportion is strong and confident, although the exterior treatment is smooth and friendly, more feminine, perhaps. The window line running along the side of the car kicks up behind the rear door, with the effect of visually adding more power to the rear shoulders while making the roof appear to slope down sportily at the rear. The sporty interior, similar to that of the Qashqai, is easy to understand and use, with clear instrumentation and switches.

Friendly, rounded and modern styling, together with an edgy name, will help Nissan to attract younger buyers with this model; unlike the Qashqai in Europe, it does not have to do double duty as a family hatchback competitor, improving its credibility as a crossover in the $20,000 market.

NISSAN

Nissan X-Trail

Design	Shiro Nakamura
Engine	2.5 in-line 4 (2.0, and 2.0 diesel, also offered)
Power	124 kW (166 bhp) @ 6000 rpm
Torque	193 Nm (142 lb. ft.) @ 4800 rpm
Gearbox	6-speed manual
Installation	Front-engined/four-wheel drive
Front suspension	MacPherson strut
Rear suspension	Multi-link
Brakes front/rear	Discs/discs
Length	4630 mm (182.3 in.)
Width	1785 mm (70.3 in.)
Height	1685 mm (66.3 in.)
Wheelbase	2630 mm (103.5 in.)
Kerb weight	1544 kg (3404 lb.)

To understand the new Nissan X-Trail it is perhaps useful to recall the surprise success of the outgoing model. Launched in 2001, its design was quite striking when first revealed; many felt the shape was uncomfortably boxy and strait-laced, but a series of good test reviews helped it to gather significant market momentum. Before long it was selling at almost twice the rate originally projected.

But as is so often the case when the time comes to renew a model that has been an unexpected hit, company planners were reluctant to depart from the successful theme for fear of losing the unknown element that was the key to success. So the second-generation X-Trail takes its cue directly from its predecessor, to the extent that they are hard to tell apart.

The new car's proportion is very similar to the outgoing model; the windscreen is if anything slightly more upright and the top of the rear screen curves forward slightly more. The rear quarter-window and high-mounted rear lamps now create a boomerang-shaped C-pillar.

Although the new X-Trail looks strong and has big features, it is friendlier and more feminine than it first appears. The large headlamps sandwich a small, friendly grille, the tall windows and low waistline give excellent visibility and do not make the occupants look overly protected, and rounded edges to the side windows and the body surfacing again add to its soft, friendly appearance.

Longer and with a bigger boot than the old model, the new X-Trail is in fact based on the platform of the Qashqai, a recent, more mainstream soft-road addition to the Nissan range. This has allowed the X-Trail to become a more focused genuine 4x4, strengthening its appeal against the new generation of soft-riding crossovers. The real test, however, will come three years down the line when the styling, substantially unchanged for a decade, will really begin to look dated.

NISSAN
X-TRAIL

Opel GTC

Design	Bryan Nesbitt and Anthony Lo
Engine	2.8 V6
Power	224 kW (300 bhp)
Torque	400 Nm (295 lb. ft.) @ 1850–4500 rpm
Gearbox	6-speed manual
Installation	Front-engined/all-wheel drive
Brakes front/rear	Discs/discs
Front tyres	245/40R20
Rear tyres	245/40R20
Length	4830 mm (190.2 in.)
Width	1867 mm (73.5 in.)
Height	1432 mm (56.4 in.)
Wheelbase	2737 mm (107.8 in.)
Track front/rear	1627/1629 mm (64.1/64.1 in.)
0–100 km/h (62 mph)	6 sec
Top speed	250 km/h (155 mph) limited

This is not the first time that Opel has presented a coupé concept called GTC at the Geneva show; in 2003 the GTC Genève previewed the style that was to become today's Astra. The mission for this rather larger coupé is, suitably, even more ambitious: this is the car that launches the fresh identity of the upcoming 2008 Vectra replacement, and with it the look of Opel's entire post-2008 line-up.

Like its competitor the Ford Mondeo, the Vectra badly needs an injection of style and glamour to restore Opel's fortunes in the large-saloon segment, where volume brands have taken a pasting from premium operators. And, again like Ford, Opel has chosen to preview the new image with a suave coupé concept rather than a sedate saloon, though Ford's Iosis (the *Car Design Yearbook 5*) was rather bolder than Opel's effort despite offering reasonable rear-seat accommodation.

In terms of its proportions the GTC is similar to the Nissan Z coupé, with a long arched roof canopy sitting on a high waistline rising between the front and rear lamps. The front lamps are unusual in that they wrap back around the fenders to link into vertical air intake slots in the front face. This, hints Opel, is the brand's new frontal signature. A well-detailed boomerang graphic features on the doors and breaks up what would otherwise be a large planar surface, while flared wheel arches push the wheels outside the body to give the car a planted stance on the road. The stacked exhaust tailpipes echo the front air intakes and are the strongest graphic element at the rear.

The GTC concept is a strong masculine design that lacks the finesse of, say, the new Ford Mondeo. It seems somewhat short on genuine innovation, and Opel will have to make it simpler and calmer if it is to translate it into a successful new look for the whole brand.

Opel/Vauxhall Antara

Design	Bryan Nesbitt
Engine	3.2 V6 (2.4 in-line 4, and 2.0 diesel, also offered)
Power	167 kW (224 bhp) @ 6600 rpm
Torque	297 Nm (219 lb. ft.) @ 3200 rpm
Gearbox	5-speed automatic
Installation	Front-engined/four-wheel drive
Front suspension	MacPherson strut
Rear suspension	Multi-link
Brakes front/rear	Discs/discs
Front tyres	215/70R16
Rear tyres	215/70R16
Length	4575 mm (180.1 in.)
Width	1850 mm (72.8 in.)
Height	1704 mm (67.1 in.)
Wheelbase	2707 mm (106.6 in.)
Track front/rear	1562/1572 mm (61.5/61.9 in.)
Kerb weight	1865 kg (4112 lb.)
0–100 km/h (62 mph)	8.8 sec
Top speed	203 km/h (126 mph)
Fuel consumption	11.6 l/100 km (24.4 mpg)
CO_2 emissions	278 g/km

Despite its European branding, the Opel Antara is built on the GM Theta platform developed by GM Daewoo in Korea, and is made in Korea, too. In fact, well-travelled readers may recognize doubles of the Opel Antara middleweight SUV around the globe. Structurally, it is closely related to the Chevrolet Captiva (also sold as the Daewoo Winstorm), while there is an even stronger panel-for-panel identification with the second-generation Saturn Vue, also featured in this book, sold in North America.

European Antaras are Vauxhall-branded in the UK but known as Opels in the rest of Europe, while in Australia the identical vehicle will be the top model of the Captiva line-up of the Holden brand, under the name Captiva MaXX. This crossover SUV was first shown as a concept in 2005 and again in the guise of the Saturn PreVue in 2006; it has now become one of the most complex global branding exercises ever undertaken by the auto industry.

The Antara has all the characteristics one would expect to see on a mid-size crossover SUV. Its fresh but not racy profile hints at an active lifestyle, although the relatively limited ground clearance suggests that any off-road use will be cautious. Designed above all as a family car, it has the high seating position many buyers currently favour. A spacious, light-filled interior helps to promote a casual, easy-going image, without sacrificing practicality.

As with the Antara's sister models, chrome accents pick out the arched profile of the window line, tapering towards the rear, and add interest to the sides, though the vertical air vent just aft of the front wheel is an incongruous detail. A bright metal skidplate is just visible under the front bumper, again alluding to the toughness of purpose that the SUV customer likes to display.

GG A 8608

Opel/Vauxhall Corsa

Design	Niels Loeb
Engine	1.7 in-line 4 diesel (1.3 diesel, and 1.0, 1.2 and 1.4 petrol, also offered)
Power	93 kW (125 bhp) @ 4000 rpm
Torque	280 Nm (206 lb. ft.) @ 2300 rpm
Gearbox	5-speed manual
Installation	Front-engined/front-wheel drive
Front suspension	MacPherson strut
Brakes front/rear	Discs/discs
Length	3999 mm (157.4 in.)
Width	1737 mm (63.4 in.)
Height	1488 mm (58.6 in.)
Wheelbase	2511 mm (98.9 in.)
Track front/rear	1485/1478 mm (58.5/58.2 in.)
Kerb weight	1160 kg (2558 lb.)

In choosing to launch its new Opel (Vauxhall in the UK) Corsa at the London Motor Show in 2006, General Motors clearly signalled its view of the UK market as an important one. Across the broader perspective of Europe, more than 9.4 million units of the popular Opel's previous models have been sold since 1982, the immediate predecessor selling steadily despite an indifferent reception from the design community. Reflecting an awareness of this, the latest version has some new and exciting style changes designed to ensure a higher visual profile and more attention from potential buyers.

In contrast to the older version, the new Corsa employs very different design themes for the three- and five-door models, the specific intention being to align each version more closely with the preferences of each customer group. The three-door is visibly more sporty and comes with muscular wheel arches and a dynamic coupé-like arched side profile. The triangular rear side window is similar to that found on the larger sports Astra GTC; the five-door, on the other hand, has an apparently higher roofline, emphasizing interior space, and a squarer rear door shape similar to that found on its family-friendly competitor the Renault Clio.

This new Corsa is quite a change from the old version, being 40 mm (1.5 in.) taller and 150 mm (6 in.) longer, and having much more energetic body-surface language. It is distinctively more modern, with both variants having a steeply raked bonnet and windscreen. It has a friendly face, too, characterized by the large headlamps and the multitude of lateral lines that rise towards the outside of the car in a 'smile'. From the rear the friendly nature continues with a series of soft rolling curves that flow across the car and take in the rear lamps.

Inside, a partially successful attempt has been made to try to improve perceived quality, a regular Opel failing. Unfortunately, mixing gloss black surfaces makes the cheap plastic look even cheaper.

GG YN 789

Peugeot 4007

Engine	2.2 in-line 4 diesel
Power	116 kW (156 bhp) @ 4000 rpm
Torque	380 Nm (280 lb. ft.) @ 2000 rpm
Gearbox	6-speed manual
Installation	Front-engined/all-wheel drive
Front suspension	MacPherson strut
Rear suspension	Multi-link
Brakes front/rear	Discs/discs
Length	4640 mm (182.7 in.)
Width	1800 mm (70.9 in.)
Height	1670 mm (65.7 in.)
Fuel consumption	7.3 l/100 km (38.7 mpg)
CO_2 emissions	194 g/k

The three French carmakers are very late in arriving at the SUV party; so late, in fact, that the party may well be over by the time the models appear in the showrooms. All three have had to resort to Asian friends to supply these models: Renault to Samsung in Korea and PSA–Peugeot Citroën to Mitsubishi, where the new Citroën C-Crosser and Peugeot 4007 are built using the Japanese company's Outlander platform and interior and much of its sheet metal too.

With its 4007 designation's double zero signifying that it is a niche model outside the mainstream Peugeot ranges (which have single zeroes), the new crossover differs only in small details from its Citroën twin. The most immediately obvious of these is the frontal styling: whereas the Citroën has an almost subtle smile emanating from its double chevron grille and broad, shapely headlights, the aggressive big-mouth treatment chosen for the Peugeot comes as quite a shock.

The full-width chrome-bordered grille is an exaggerated extension of Peugeot's current brand identity theme. It gives the 4007 the air of a giant predatory animal; the vertical chromed slats resemble teeth and the black rubber-faced bumper set in the centre could be the remains of an unfortunate smaller vehicle that the 4007 has just gobbled up. In addition, the Peugeot lion logo seems unnaturally high on the bonnet and the triangular headlamps look rather menacing. It is almost as if Peugeot is trying to outdo the very expensive Porsche Cayenne for aggressive ugliness.

In other respects the 4007 is bolt-for-bolt identical to Citroën's C-Crosser; as such, it is likely to have a harder time convincing buyers already worried by concerns surrounding SUVs. The first Peugeot to have a double-zero designation, the 1007 of 2005, has been a poor seller: now, with the bigger 4007's over-aggressive bearing, Peugeot could have compromised its niche chances a second time.

37 DSG 78
PEUGEOT

Peugeot 908 RC

Design	Gerald Welter and Amko Leenarts
Engine	5.5 V12 diesel
Power	522 kW (700 bhp)
Torque	1200 Nm (884 lb. ft.)
Gearbox	6-speed automatic
Installation	Mid-engined/rear-wheel drive
Front suspension	Double wishbone
Rear suspension	Double wishbone
Brakes front/rear	Discs/discs
Front tyres	255/35R20
Rear tyres	285/30R21
Length	5123 mm (201.7 in.)
Height	1370 mm (53.9 in.)
Wheelbase	3150 mm (124 in.)
Kerb weight	1800 kg (3968 lb.)
Top speed	299 km/h (186 mph)

The Peugeot 908 RC concept is flamboyant in its shaping and dramatic in its proportions, as it inevitably must be, given the fact that it attempts to do the almost impossible in packaging a four-door, four-seater luxury cabin into the mid-engined layout normally reserved for such extreme supersports cars as the Bugatti Veyron.

Peugeot has dallied with the mid-engined saloon concept in several previous studies, none of which yielded any production-car carry-over. Though the imposing 908 is spectacular, it is equally unlikely to influence mainstream models; its main role is to showcase the huge V12 diesel engine with which Peugeot is contesting the Le Mans 24 Hours endurance race in 2007.

The position of the engine enforces a very cab-forward seating position, with the short bonnet truncating less than a metre from the windscreen. The screen itself is huge, resulting in a header rail that is moved back into line with the B-pillar to create a sense of freedom from the inside.

Peugeot has sought to communicate its lion emblem in the 908's design language through a snarling cat-like face and pronounced haunches over the rear wheels. The side view shows the car to be almost symmetrical front to back; the front design is hugely curvaceous with a complex interplay of lines where the fenders and lights meet the humped centre part of the bonnet. The rear end, with its angular planar surface, looks severe and intimidating in its bid to reconcile supercar and luxury-car cues.

The interior makes much use of stitched leather in sweeping, sporty-looking forms. In contrast to the exterior, it is smooth and simple, with the very minimum of buttons and controls.

This is clearly an evocative car that may well grab glossy picture space in the post-Paris show reviews. Yet, once again, it is a Peugeot concept that appears to have little relevance to the company's everyday output.

908 RC

Pontiac G8

Engine	6.0 V8 (3.6 V6 also offered)
Power	270 kW (362 bhp) @ 5700 rpm
Torque	531 Nm (391 lb. ft.) @ 4400 rpm
Gearbox	6-speed automatic
Installation	Front-engined/rear-wheel drive
Front suspension	MacPherson strut
Rear suspension	Multi-link
Brakes front/rear	Discs/discs
Front tyres	245/45R18
Rear tyres	245/45R18
Length	4981 mm (196.1 in.)
Width	1980 mm (74.8 in.)
Height	1466 mm (57.7 in.)
Wheelbase	2916 mm (114.8 in.)
Track front/rear	1593/1608 mm (62.7/63.3 in.)
Kerb weight	1812 kg (3995 lb.)

Pontiac's new G8 replaces the Grand Prix, a long-running Pontiac nameplate that had many ups and downs over the past decades; in its final form, the model lacked the necessary personality to be noticed by the US public. The new G8 taps into the sporty part of the Grand Prix tradition and is emphatically a sports sedan, as evidenced by its rash of grilles, slots, air scoops and aerodynamic aids.

Impressive though this muscular body kit undoubtedly is, the G8 is in fact a rebadged version of Australia's Holden Commodore with the appropriate cosmetic tweaks – including, of course, the trademark Pontiac divided grille, in this instance with chrome frames edging the black wire mesh. Built on GM's Zeta platform, the G8 brings a much more powerful model to Pontiac's line-up, with a huge 6-litre V8 option listed. This platform is of course rear-wheel drive, a key driver demand in the high-horsepower sedan stakes.

Pontiac's performance aspirations are easily seen in the design by the multiple frontal air intakes, the twin air scoops in the bonnet, and the piercing circular lenses mounted within the headlamp housings. The body has well-defined features, including sharp creases in the bonnet and strong deep sills, while the large splayed wheel arches push the wheels right out to the edge of the body to make the car appear well planted on the ground. Yet it is those four tailpipes at the rear that are the biggest exterior sign of the car's power.

The G8 is a car that divides opinions, not so much in relation to its styling, which all concede is necessarily aggressive, but on the question of whether GM should be fostering its own home-grown high-performance talent rather than going for the quick-fix remedy of an off-the-shelf solution from one of its remotest affiliates.

Renault Koleos

Design	Patrick le Quément
Engine	2.0 in-line 4 diesel
Power	134 kW (180 bhp)
Gearbox	6-speed manual
Installation	Front-engined/four-wheel drive
Brakes front/rear	Discs/discs
Front tyres	255/50R19
Rear tyres	255/50R19
Length	4520 mm (178 in.)
Width	1890 mm (74.4 in.)
Height	1700 mm (66.9 in.)

The Koleos name might be familiar, and so, too, might be the idea of a Renault SUV show concept. But this vehicle is radically different from the ambitious and much-praised concept of the same name, first aired in 2000: today's Koleos is an almost production-ready crossover SUV and its smart, though restrained, lines lack the drama of the original's daring elliptical side shaping.

The explanation for this 180-degree turn lies in the complex politics – and economics – of platform sharing within the Renault–Nissan alliance, which also includes Samsung of Korea. When production of the Koleos begins in late 2007 it will be in South Korea, and Samsung will market a very similar vehicle at the same time; additionally, both models incorporate off-road engineering expertise from Nissan, meaning some crossover thinking with that company's new Qashqai, too.

The Koleos comes across as less playful than other Renault models, yet it is undeniably an attractive-looking car, with a rounded and friendly face. In this, it is the polar opposite of the very aggressive Ford Iosis X, another of this year's crossover SUV concepts. The DLO is gently arched, rather than wedge-shaped, and the relationship of the C-pillar to the side shaping is particularly effective.

The rear end is softened by such small details as the screen encasing the badge, and the rear lamps having an arc at their base; these are small things that make a huge difference. The rectangular aluminium bar in the rear lamps and the swagged lower stone-guards around the base of the car are overly fussy by comparison.

The interior is friendly in a small-car sort of way, using small hooded instruments and a combination of dark browns and creams to create a modern look. It, too, appears almost production-ready, but Renault might be well advised to soften up the colourways to give the passenger compartment an even greater level of friendliness, to match that of the exterior.

KOLEOS
KOLEOS
Concept

Renault Nepta

Design	Patrick le Quément
Engine	3.5 V6
Power	313 kW (420 bhp)
Torque	560 Nm (413 lb. ft.) @ 3000 rpm
Gearbox	7-speed automatic
Installation	Front-engined/rear-wheel drive
Brakes front/rear	Discs/discs
Front tyres	R23 PAX 275
Rear tyres	R23 PAX 275
Length	4995 mm (196.7 in.)
Width	1956 mm (77 in.)
Height	1332 mm (52.4 in.)
0–100 km/h (62 mph)	4.9 sec

Anyone could be excused for thinking that the Nepta is a fanciful dream from Renault; after all, such hugely esoteric features as its gullwing doors mark it out as a design that is miles away from the current Renault model line-up. But then Renault has openly signalled its desire to extend its offerings upmarket, and a number of concept cars in the recent past – such as the Talisman and the Fluence – have also supported this.

Changing the perception and positioning of such a well-established brand as Renault needs to be done with great care and gently over time, so that customers hardly realize it is happening. As part of this strategic evolution, the Nepta might even be seen as the inspiration for a market competitor to such classy convertibles as the Audi A4 or Volvo C70.

The trouble is, the Nepta's doors are so fanciful that it is hard to know how serious Renault is about this concept. On a production car it would be next to impossible to engineer the structure, which sees almost the whole length of the car hinge up from the central longitudinal axis. And there is no sign of a roof solution, either.

The Nepta has a smooth aerodynamic form made up of long surfaces that gracefully wind themselves from front to rear. It is classy, elegant and luxurious in every detail. The boot lid gradually slopes down like a boat-tail and the long, thin rear lamps help to accentuate the long, slender body. The quality of such a complex concept car is to be admired in itself: it is in these exciting and unique designs that Renault truly excels. The interior, for example, is kept simple and uses only high-quality red, grey and white leather in combination with complex aluminium instrumentation and controls.

Renault Twingo

Design	Patrick le Quément
Engine	1.2 in-line 4 (1.5 diesel also offered)
Power	75 kW (100 bhp) @ 5500 rpm
Torque	145 Nm (107 lb. ft.) @ 3000 rpm
Gearbox	5-speed manual
Installation	Front-engined/front-wheel drive
Front suspension	MacPherson strut
Rear suspension	H-beam
Brakes front/rear	Discs/drums
Length	3600 mm (141.7 in.)
Width	1654 mm (65.1 in.)
Height	1470 mm (57.9 in.)
Wheelbase	2367 mm (93.2 in.)
Track front/rear	1414/1400 mm (55.7/55.1 in.)
0–100 km/h (62 mph)	9.8 sec
Fuel consumption	5.9 l/100 km (47.9 mpg)
CO_2 emissions	140 g/km

Some might say that Renault missed a trick by waiting so long to replace its cutesy, charming Twingo model, originally launched in 1993. With such competitors as the Citroën C1 and the Peugeot 107 snapping at its heels, and with the political shift towards smaller cars intensifying, Renault could hardly have waited much longer. But perhaps the biggest missed trick is to replace such a charismatic and much-loved vehicle with a model that, on the surface at least, appears to be anonymous and to be following trends rather than setting them.

While the new Twingo does share a certain simplicity evident in the original, it lacks the distinction that, even fourteen years on, still marks out that first model. Instead, it synthesizes modern market-driven features and reflects Renault's latest design trends.

For example, the rear wheels are positioned far rearward, pushing the bumper out, as on the larger Modus. The rear screen drops in the centre, pointing to the Renault lozenge badge. The all-important Twingo face is only just visible in the large headlamps and the rounded and simplistic front end. There is certainly no smile.

Instead, and in order to woo a youthful clientele, most of the attention has gone into the interior. The Twingo can claim iPod and smartphone inputs, a USB port, and graphic-equalizer-style dashboard decoration; there is even the option of an Internet connection, webcam and monitor screen, and individual plug sockets for every seat.

The interior design is certainly more fun, with a two-tone dashboard, a rev-counter mounted behind the steering wheel, and behind that a neat spiky rubber contraption that holds various bits and pieces. But no amount of digital trickery can compensate for the personality bypass the new car appears to have suffered; it may be highly competent technically, but from Renault we expect imagination not imitation. The original lasted fourteen years without any change; this one will be lucky to manage four.

TWINGO
RENAULT

Rinspeed eXasis

Design	Rinspeed
Engine	0.75 two cylinder bioethanol supercharged
Power	112 kW (150 bhp) @ 7500 rpm
Torque	150 Nm (111 lb. ft.) @ 4500 rpm
Gearbox	6-speed manual
Installation	Rear-engined/rear-wheel drive
Front suspension	Double wishbone
Rear suspension	Double wishbone
Brakes front/rear	Discs/discs
Front tyres	265/30R22
Rear tyres	295/25R22
Length	3700 mm (145.7 in.)
Width	1960 mm (77.2 in.)
Height	1284 mm (50.6 in.)
Wheelbase	2500 mm (98.4 in.)
Track front/rear	1700/1700 mm (66.9/66.9 in.)
Kerb weight	750 kg (1653 lb.)
0–100 km/h (62 mph)	4.8 sec
Top speed	210 km/h (130 mph)

One of the most outlandish concepts shown at Geneva, Rinspeed's eXasis is a collaboration with Bayer MaterialScience AG, and aims to demonstrate what is possible in plastics technology. As such, it is something of a Bayer materials brochure on four wheels.

The eXasis's most visible – or perhaps that should be invisible – feature is its yellow-tinted transparent body, made from Bayer Makrolon polycarbonate, wrapping around the aluminium chassis. The lightweight plastic that encloses the car not only channels the airflow around the body and protects the driver and passenger, but also allows the technical rawness of the car to remain visible.

The layout is like an older-style racing car with a cigar-like fuselage and a tandem seating arrangement, to ensure minimum frontal area; suspension is racing car like, too, and the tiny engine is behind the passenger.

The interior provides another handy product listing for Bayer. The transparent instrument and function displays are made from the CD material Makrolon and, claims Rinspeed, appear to hover on both sides in the driver's field of vision; they are coated with electrically conductive Baytron so that by touching them various functions can be displayed and controlled. Transparent Technogel is used for head restraints in the Recaro-designed deckchair-like seats and on the armrests; the seats adapt themselves to the shape of the occupants.

A standard steering wheel would be far too conventional for this eccentric vehicle: the handlebar-like device is finished in a yellow metal lattice, perhaps appropriate for such a minimal machine design to give the driver maximum exhilaration and connection with the road. And for anyone dubious about the translucent finish, just remember the original iMac and how it led to a rash of imitators across every area of consumer electronics.

RINSPEED
eXASIS
Weber Motor
PIRELLI

Weber Motor
makrolon
BioPower
MECAPLEX
KABA
alcosuisse

Saturn Aura

Design	Clay Dean
Engine	3.6 V6 (2.4 in-line 4 hybrid and 3.5 V6 also offered)
Power	188 kW (252 bhp) @ 6400 rpm
Torque	340 Nm (251 lb. ft.) @ 3200 rpm
Gearbox	6-speed automatic
Installation	Front-engined/front-wheel drive
Front suspension	MacPherson strut
Rear suspension	Multi-link
Brakes front/rear	Discs/discs
Front tyres	225/50R18
Rear tyres	225/50R18
Length	4851 mm (191 in.)
Width	1786 mm (70.3 in.)
Height	1464 mm (57.6 in.)
Wheelbase	2852 mm (112.3 in.)
Track front/rear	1523/1533 mm (60/60.4 in.)
Kerb weight	1654 kg (3647 lb.)
Fuel consumption	9 l/100 km (31 mpg)

The Saturn Aura could hardly be more conventional: it is a mid-size saloon, replacing the fast-fading L-series in the North American market. But to Europeans it is very familiar indeed, for, in a bid to avoid later slack sales, GM's US-only Saturn brand has turned to Germany's Opel – also GM-owned – and its mid-market Vectra for inspiration.

More than just inspiration, in fact: the concept version of the Aura, shown just over a year before the production model was launched, was very closely related to the European product, and, save for a few minor detailed tweaks, the final design is identical. So Saturn is presenting a well-proven product.

The detailed changes to the Vectra's styling for the Aura reflect Saturn's new design language. Particular points include the now-common jewel-like headlamp theme and the chrome grille bar bearing the red Saturn logo. The rear lamps are much more emphatic than on the European model, with a thin solid bar encircling them and a horizontal line through the red reflector lens.

The long, gently sweeping roofline and, especially, the long rear side window lend the Aura a premium appearance, emphasizing passenger comfort. The chrome trim encircling the side windows also seeks to add to the premium feel.

The more powerful end of Saturn's engine line-up is expected to be complemented by a 2.4-litre Ecotec-based hybrid, similar to that used in the Vue Green Line SUV. It is GM's first hybrid passenger car and provides a telltale sign that the market is moving towards greener options.

Inside, the layout is easy to use and, again, the feel is very much that of the European Vectra, albeit with the colouring of brown leather mixed with grey plastic clearly tailored for American tastes.

Saturn Outlook

Design	Clay Dean
Engine	3.6 V6
Power	205 kW (275 bhp) @ 6600 rpm
Torque	340 Nm (251 lb. ft.) @ 3200 rpm
Gearbox	6-speed automatic
Installation	Front-engined/all-wheel drive
Front suspension	MacPherson strut
Rear suspension	H-arm independent
Brakes front/rear	Discs/discs
Front tyres	255/65R18
Rear tyres	255/65R18
Length	5097 mm (200.7 in.)
Width	1986 mm (78.2 in.)
Height	1846 mm (72.7 in.)
Wheelbase	3020 mm (118.9 in.)
Track front/rear	1704/1704 mm (67.1/67.1 in.)
Kerb weight	2239 kg (4936 lb.)
Fuel consumption	11 l/100 km (26 mpg)

In replacing the big and boxy Relay minivan, sales of which had been struggling, Saturn's designers clearly hope that the switch to a more progressive design will change the fortunes of the brand in the people-carrier segment.

Accordingly, the Outlook, built on the GM Lambda platform that yielded the Buick Enclave of 2006 (see the *Car Design Yearbook 5*), is a much more expressive design than the utilitarian Relay: the new car's shape is more compact and curvaceous, though without the flowing curves and pronounced wheel arches of the Enclave. Instead, the wheel arches are squared off in an interesting manner, an effect enhanced by the contrast colour of the lower body. The rear section behind the wide, body-coloured C-pillar looks visually separated from the main passenger space by the wraparound rear screen that tucks up to the roof and breaks up the side view.

The front incorporates a bolder iteration of Saturn's new facial identity, with a deeper grille carrying the company logo on a bright central bar, and strong headlamps set back several inches from the line of the bumper and the protruding grille. Other features help to project a premium image too: the chrome-rimmed tail lamps, the chrome door handles, a small roof spoiler and smart aluminium roof rails.

The Outlook seats up to eight people, with an industry-first 'SmartSlide' system that gives easier access to the three-seat rearmost bench as the second-row seats automatically slide forward and their seat cushions flip and compress. This layout contrasts with the six separate seats in the Buick's interior.

From being a strictly one-model brand for many years, and suffering in the marketplace as a result, Saturn at last has a broad product line-up and a strong design identity with which to compete against the US domestics and the Asian automakers in the North American market.

California
5NNJ983
OUTLOOK XR
AWD

Saturn Vue

Engine	3.6 V6 (2.4 in-line 4 and 3.5 V6 also offered)
Power	186 kW (250 bhp) @ 6500 rpm
Torque	329 Nm (242 lb. ft.) @ 4400 rpm
Gearbox	5-speed manual
Installation	Front-engined/all-wheel drive
Front suspension	MacPherson strut
Rear suspension	Multi-link
Brakes front/rear	Discs/discs
Front tyres	235/60R17
Rear tyres	235/60R17
Length	4576 mm (180.2 in.)
Width	1850 mm (72.8 in.)
Height	1704 mm (67.1 in.)
Wheelbase	2707 mm (106.6 in.)
Track front/rear	1562/1572 mm (61.5/61.9 in.)
Kerb weight	1962 kg (4325 lb.)

It may be an all-new Saturn for the 2008 model year, but there is already something distinctly familiar about this latest version of the Vue. The answer lies in General Motors' global strategy: all the group's middleweight SUVs are based on a common platform developed and built by Daewoo in Korea. The idea is to maximize sales of each platform and thus gain economies of scale. And that in turn is why this model is already on sale, with only minor external differences, as the Chevrolet Captiva and the Opel Antara in Europe, as well as under a variety of other labels in other continents.

As with the Antara and the Captiva, the Vue's design is smooth and superficially sporty, with a rounded nose and an angled-forward tailgate. Brand differentiation for the Saturn model is confined to the substitution of the red Saturn logo for the Opel 'Blitz' symbol on the centre of the wide chrome band running across the grille. Chrome accents are evident elsewhere, such as around the side windows and at the base of the tailgate; these are common to the Opel, too, but the Saturn's rear lights forgo the distinctive clear circular centre sections of the European version. A tight wheel-to-body relationship and the closeness to the ground make it obvious that this is a soft-roader only; European models are perhaps more pretentious in this respect, with their tough-looking dark-grey bumpers, side cladding and wheel-arch surrounds. On the Saturn these are all body-coloured.

Inside, there are minor differences in the shaping of the centre console; little else is altered. As in all Saturn vehicles, the interactive OnStar safety and security communications system is included.

AM
FM
XM
DISC
AUX
FM 93.5

Scion Fuse

Scion is Toyota's North American youth brand and regularly comes up with concepts and production cars that cross boundaries to challenge existing ideas. The bold Fuse is just such a design. Curvy, coupé-like style cues compete with razor-sharp features to create an energized, testosterone-fuelled car with dramatic effect.

Butterfly doors are a sure-fire attention-grabber for the Generation-Y market that Scion is aiming for, and the mean-looking side profile has narrow, lozenge-shaped windows, a feature inspired by racing helmets to create a feeling of secure encapsulation for the occupants. The shape is also featured on the front lamps to create an equally mean-looking front end. The wide-open rectangular jaw formed by the grille is suggestive of the power that lurks within, as is the narrow air intake slot at the forward edge of the bonnet. The high waistline allies with the helmet-like C-pillar to give a strong sense of power and protection, while the tyre-hugging wheel arches and low sills visually ground the car to suggest high powers of roadholding.

An extremely innovative concept for the indicators are LEDs set at the end of each wheel spoke. Personalization is important to the young Scion customer: lamps can be selected to illuminate in various colours and, more significantly, the interior – which initially looks like a driver-focused single-seat production racer – can transform itself into an entertainment centre for watching movies and playing games or, by a quick fold of the front passenger seat, can provide a table and allow friends to join in too. There's even a bench that slides out of the cargo floor to seat two.

From the outside, the Fuse is not exactly a friendly-looking coupé, but then it was never intended to be. Instead, it is an extension of US street culture, showing how keen Scion is to capitalize on an area of demand that is currently served by aftermarket customizers rather than the established automakers.

Design	Alex Shen and Bob Mochizuki (exterior); Alan Schneider and Ichirou Mukai (interior)
Engine	2.4 in-line 4
Installation	Front-engined/front-wheel drive
Brakes front/rear	Discs/discs
Front tyres	245/35R20
Rear tyres	245/35R20
Length	4420 mm (174 in.)
Width	1796 mm (70.7 in.)
Height	1323 mm (52.1 in.)
Wheelbase	2700 mm (106.3 in.)

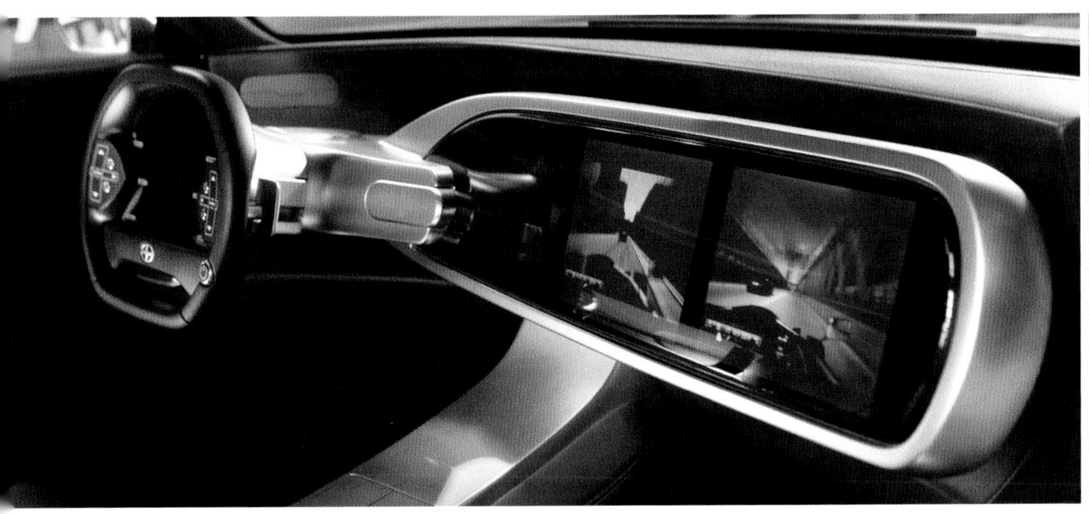

scion

Scion xB

Engine	2.4 in-line 4
Power	118 kW (158 bhp) @ 6000 rpm
Torque	220 Nm (162 lb. ft.) @ 4000 rpm
Installation	Front-engined/front-wheel drive
Front suspension	MacPherson strut
Rear suspension	Torsion beam
Brakes front/rear	Discs/discs
Front tyres	205/55R16
Rear tyres	205/55R16
Length	4249 mm (167.3 in.)
Width	1760 mm (69.3 in.)
Height	1590 mm (62.6 in.)
Wheelbase	2601 mm (102.4 in.)

Scion, the youth brand under Toyota, unveiled its all-new 2008 xB urban utility vehicle at the Chicago Auto Show in 2007. The original xB's boxy shape quickly came to be seen as hip and trendy among streetwise young buyers and soon became the defining style for the still-young Scion brand. Responding rapidly to the results of research conducted among its customer body, Scion has moved to produce an upgraded and enlarged version of the xB.

The new car is indeed significantly larger: it has grown by 30 cm (12 in.) in length and gains a major capacity and power hike, now using the same 2.4-litre engine as is fitted to the tC coupé, offering 55 horsepower more than the previous 1.5-litre unit. Although the new model retains the characteristic boxiness, it has a subtly different proportion and is slightly lower and narrower than the original. Its lower centre of gravity is visually helped by the high waistline that is retained to give a protective appearance. The rear screen now angles slightly forward, as does the very large and solid C-pillar. As before, the car is comprised of large, solid flat panels rounded off at the edges to soften the effect; if anything the new car is less severe in its appearance than the outgoing xB.

The interior design features instrumentation offset from the centre line, the idea being to share information with passengers; below the line of four overlapping orange instruments is the 160-watt audio system panel, offset further to the passenger side. Perhaps oddly for such an interactive, youth-orientated car, there is no central display screen to allow images or information, such as iPod track listings, to be shared among friends.

The new xB is without question a more focused car to fit with the needs of America's youth culture, but apart from the distinctive exterior, it is still rather tame in terms of innovation and features.

DISP
ODO-TRIP
BRAKE

Scion xD

Engine	1.8 in-line 4
Power	95 kW (128 bhp) @ 6000 rpm
Torque	170 Nm (125 lb. ft.) @ 4400 rpm
Gearbox	5-speed manual
Installation	Front-engined/front-wheel drive
Front suspension	MacPherson strut
Rear suspension	Torsion beam
Brakes front/rear	Discs/drums
Front tyres	195/60R16
Rear tyres	195/60R16
Length	3929 mm (154.7 in.)
Width	1725 mm (67.9 in.)
Height	1524 mm (60 in.)
Wheelbase	2461 mm (96.9 in.)

Replacing Scion's somewhat anonymous and undistinguished xA, the new xD is much more in tune with the Scion-defining spirit of the fresh xB, also featured in this edition of the *Car Design Yearbook*.

The original xA always seemed much more like a conventional European hatchback, whereas the xB had a strong youth identity, the strength and success of which in turn shaped the public image of the whole Scion brand. The new xD retains the fundamental European hatchback proportions of the old xA but is now much more tough-looking and packed with attitude.

Based on the Toyota Yaris platform, the xD shares with its xB brother the high waistline, flattish hood line, strong C-pillar, horizontal headlamps and deep front bumper. Yet, proportionally, the cars are still quite different: the xD has an arched roofline (inherited from the Yaris's structural hard points) and is formulated from curves and rolling surfaces in contrast to the xB's planar and boxy elements.

The interior is very different from the xB as it features a more conventional, driver-focused instrumentation layout, with a single gauge combining the tachometer and speedometer mounted behind the steering wheel. The interior finishes are interesting, although to European eyes the effect is not high quality. Glossy black panels adorn the dashboard, as do multiple curved surfaces, which all add up to an energy-fuelled design that is certainly not relaxing. As with the xB, apart from a premium stereo, there are no special new technologies to appeal to the youth market of this digital age.

Indeed, it is as if the xD had been conceived as the car for those young people who like the idea of being associated with urban street culture via the Scion brand, but who do not have the confidence to opt for the more outgoing xB.

Skoda Fabia

Design	Jens Manske
Engine	1.6 in-line 4 (1.2 and 1.4, and 1.4 and 1.9 diesels, also offered)
Gearbox	6-speed Tiptronic
Installation	Front-engined/front-wheel drive
Front suspension	MacPherson strut
Rear suspension	Semi-independent
Brakes front/rear	Discs/discs
Length	3992 mm (157.2 in.)
Width	1642 mm (64.6 in.)
Height	1498 mm (59 in.)
Fuel consumption	7.5 l/100 km (37.7 mpg)
CO_2 emissions	180 g/km

The new Fabia is more evolution than revolution. Based on the same platform as the previous model, it has resisted the general temptation to expand in size; instead, it has grown only in height, but has also been noticeably updated and upgraded to reflect the latest Skoda design thinking.

The front end has a more rounded shape, with an inflated bonnet and more curvaceous headlamps. The large Skoda badge on the chrome radiator crossbar is similar to that found on the Roomster; indeed, the whole front is similar. The bonnet surface rolls over at the side into the fenders, setting up a shoulder line that twists as it goes rearward along the car until it eventually flows into the top surface of the rear lamps. The upper A- and B-pillars are blackened to give the glazing a more wraparound look, something again found on the Roomster. The door surfaces now curve outward, reinforcing the impression of a neat contemporary package. The rear lights are larger than before and are positioned higher up on the rear corners so as to make them more visible and to give the car a more confident look.

The interior uses a two-tone dashboard to push much of the switchgear up high in the driver's line of sight. What impresses for a small, low-cost car is the solidity of the interior trim and fittings; quality and feel are at Volkswagen levels. A range of equipment options, including satellite navigation, is available, each relating to the specification chosen.

The new Fabia may give a passing nod to the Mini and the Suzuki Swift in its contrast-colour roof and its personalization potential, but in reality it is rather different: simple, honest, easy-going and very well put together – a winner on all fronts.

119
5S8 3947

Skoda Joyster

Design	Jens Manske

Having successfully positioned itself as the brand of choice for buyers seeking straightforward practicality and quality, Skoda is now seeking to add a younger dimension with a much stronger element of fun and adventure. The 2006 Roomster mini-wagon (see the *Car Design Yearbook 5*) was the first step in this strategy; now the Joyster, also previewed as a show concept, goes one step further and more specifically targets a younger market.

The first car designed under the leadership of Skoda's new design director, Jens Manske, the Joyster floats ideas for a coupé majoring on practicality rather than sportiness. Featuring a Saab-like wraparound windscreen similar to that shown on the Skoda Roomster concept of 2003, the Joyster has a front where a great many shapes and details compete with one another, the wide grille and futuristic light clusters meeting to add visual width to the car.

The central bonnet bulge, although clearly part of the brand's identity, is more exaggerated than on production Skodas. With all except the rearmost pillars blackened, the bright yellow roof appears to float like an aerofoil above the car. The rear end is the most dramatic, with a white-and-red sandwich of lights forming a wide U-shape across the back of the car. A novel two-seat bench folds out of the back, just as on Scion's Fuse.

This is an energetic design with an interior to match. Black, yellow and satin aluminium mix to great sporting effect. The instrumentation and dashboard controls are deliberately kept very simple: Skoda designers believe that electronic technology advances so quickly that users will wish to bring in their latest devices from outside, rather than relying on built-in equipment that could soon be outdated.

Skoda is proving to be one of the most exciting of all brands to watch for, showing the potential to bring some of the most innovative new cars to our streets in the near future.

Smart Fortwo

Design	Hartmut Sinkwitz
Engine	1.0 in-line 3 (0.8 diesel also offered)
Power	63 kW (84 bhp) @ 5250 rpm
Torque	120 Nm (88 lb. ft.) @ 3250 rpm
Gearbox	5-speed automated manual
Installation	Rear-engined/rear-wheel drive
Front suspension	MacPherson strut
Rear suspension	DeDion rear axle
Brakes front/rear	Discs/drums
Front tyres	175/55R15
Rear tyres	195/50R15
Length	2695 mm (106 in.)
Width	1559 mm (61.4 in.)
Height	1542 mm (60.7 in.)
Wheelbase	1867 mm (73.5 in.)
Track front/rear	1283/1385 mm (50.5/54.5 in.)
Kerb weight	770 kg (1698 lb.)
0–100 km/h (62 mph)	10.9 sec
Top speed	145 km/h (90 mph)
Fuel consumption	4.9 l/100 km (57.6 mpg)
CO_2 emissions	116 g/km

It is a rare event in the automotive world to see a design classic that truly sets new trends and makes the industry rethink its attitude to transportation. But that is what the Smart did back in 1998, and since then 770,000 have been sold. Yet attempts to broaden the appeal of the brand have failed, despite the Fortwo's great success: the Forfour supermini and the Roadster are both now axed, and the Formore mini-SUV was cancelled before it even reached production.

At last, however, things have settled down. Smart has been integrated more closely into the parent DaimlerChrysler organization, engineers have been busy assessing customer feedback on how the little car could be improved, and the second-generation Smart is on the streets. The new car is a bit longer, at 2695 mm (106 in.), sacrificing some ease in end-on parking in order to comply with US safety regulations, and a new set of engines and transmissions addresses the most frequent complaint, that of slow gearshifts.

The design brief was a tough one: the team had to keep the same essential appeal of the car, yet also make it different. In particular, they needed to keep that cheeky *joie de vivre*, while making the car more functional, more comfortable and safer. At first glance, the new car is hard to tell from the old: projector headlamps add a high-tech touch, the door handle is now horizontal, and there are twin rather than triple rear lights each side. The hallmark Tridion safety cell is still there, though slimmer, and better-defined wheel arches and a broader tailgate make the Fortwo look more stable on the road.

The interior is more formal and serious, too, but the real success of the refresh is that the transition is hard to spot: the Smart still retains its most precious asset, the irresistible impish personality that has made it such a success among buyers.

Stola Coupé

Design	Stola
Engine	6.3 V12
Power	559 kW (750 bhp)
Torque	1360 Nm (1002 lb. ft.)
Installation	Front-engined/rear-wheel drive
Brakes front/rear	Discs/discs
Front tyres	315/25X22
Rear tyres	315/25X22
Top speed	322 km/h (200 mph)

For total, indulgent stylistic excess, look no further than this car. Built by Turin coachbuilder Stola, which produces prototype cars for automakers and one-off specials for wealthy clients, the show Coupé is based on the Brabus-tuned Mercedes-Benz S600. It will, says Stola, be made in a limited run of twenty-five bespoke examples, with customers free to fine-tune the specification to their own tastes.

The best that can be said about the Stola Coupé is that it is a bold, eccentric concept with exaggerated proportions, which perhaps belongs in the Stola museum or in a sci-fi movie rather than on the road. It is of course possible that there are clients who desire this type of extreme design and who have the means to commission one-off creations, and this concept is suitably far-out to guarantee lurid headlines that will be noticed by this type of customer.

The bonnet is immensely long and triangulated to a large bold chrome cap on the top of the grille at the front; slatted eyebrows top the blackened light housings either side. The deep front bumper has aggressive air intakes, yet the upper architecture is disproportionately tiny, resembling that from an Audi TT mounted to a barge. The rear extends way out back; proportionally, it is a bit like a grotesquely elongated Porsche 911.

Founded in 1919, Stola still retains a highly skilled staff, demonstrated by recent creations for other marques, including the Maybach Exelero, the Fisker Tramonto, and the construction of the Alfa Romeo 8C Competizione and the Citroën C-Métisse concept cars. True to bespoke-car manufacturing, Stola is inviting its customers to specify their own interiors and exterior finishes. Clients will also be able to name the car themselves, hence the working title of Stola Coupé. With the world's super-rich out to impress – though not necessarily with good taste – Stola might well be on to something.

RENAULT
RENAULT
STOLA
STAMPI
4
S.G.
STOLA

Suzuki Splash

Engine	1.2 in-line 4
Power	92 kW (123 bhp) @ 6800 rpm
Torque	148 Nm (109 lb. ft.) @ 4800 rpm
Gearbox	5-speed manual
Installation	Front-engined/front-wheel drive
Brakes front/rear	Discs/discs
Length	3780 mm (148.8 in.)
Width	1780 mm (70.1 in.)
Height	1650 mm (65 in.)

Suzuki certainly attracted attention when it first unveiled its Splash concept in Paris in 2006. A taller and roomier five-seater based on the platform of the successful Swift, the Splash could in the future form the template of a model to replace the dull and boxy Wagon R micro-minivan.

The Splash is as bold as the Wagon R is conservative. In its youthful appearance it resembles the Citroën C1, with playful design language, extensive use of glass, and wheels pushed right to the corners to give an impression of stability. When the car is viewed from the front, the large, upright bright-metal grille block, with deep air scoops to either side, comes as something of a surprise. The headlamps are strangely unnerving, with coloured LEDs. Unusually, the frontal design positively accentuates the higher bonnet and makes the front end look narrow.

Yet modern design, with a glass roof and aluminium cant rails, gives the Splash a contemporary visual lightness up top. The body colour wraps over the roof to give a roll-hoop effect, while the U-shaped LED lamps at the back are strikingly modern, especially for a small car, and would be a key recognition point on any future production version.

The interior has a blue colour scheme and is very simple and uncluttered, conveying a sense of spaciousness. It incorporates a single large speedometer with information display, and a navigation screen in the centre console that directly shows images captured by the rear-view cameras.

Some final refinement may still be necessary before production, especially around the too-extreme nose, but this is another promising design from Suzuki. With a name like Splash, it needs to project a certain sense of fun, which it does to good effect.

SPLASH

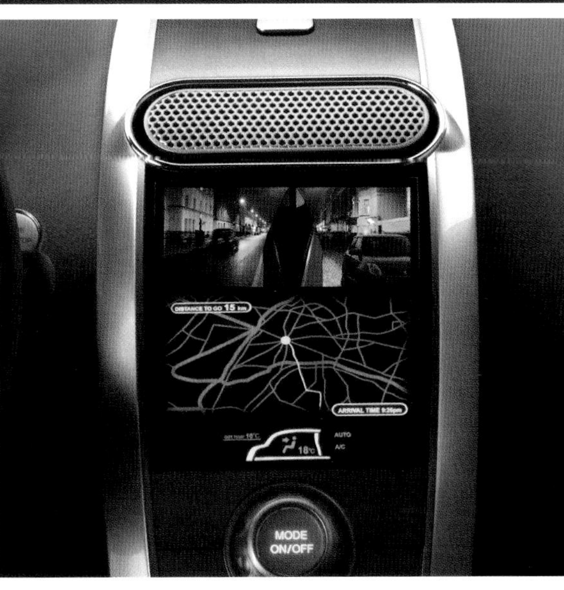
MODE
ON/OFF

0
20
40
60
80
100
120
140
160
180
SPLASH
SUZUKI
km/h

Suzuki XL-7

Engine	3.6 V6
Power	188 kW (252 bhp) @ 6500 rpm
Torque	330 Nm (243 lb. ft.) @ 2300 rpm
Gearbox	5-speed automatic
Installation	Front-engined/front-wheel drive
Front suspension	MacPherson strut
Rear suspension	Multi-link
Brakes front/rear	Discs/discs
Front tyres	235/60R17
Rear tyres	235/60R17
Length	5008 mm (197.2 in.)
Width	1835 mm (72.2 in.)
Height	1750 mm (68.9 in.)
Wheelbase	2857 mm (112.5 in.)
Track front/rear	1565/1570 mm (61.6/61.8 in.)
Kerb weight	1837 kg (4050 lb.)
Fuel consumption	11.3 l/100 km (25 mpg)

As its name implies, the new Suzuki XL-7 is considerably larger than the model it replaces, and it comes with the option of seating for seven. The first generation XL-7 was essentially a stretched Grand Vitara, using the Japanese company's own selectable four-wheel-drive platform but with a 2.7-litre V6 engine fitted. The new car, by way of contrast, is based on a larger front-wheel-drive platform, General Motors' Theta architecture. It will be built alongside the Chevrolet Equinox and Pontiac Torrent at CAMI Automotive in Ontario, Canada.

The new car was first seen in the form of the radical Suzuki Concept X at the 2005 North American International Auto Show in Detroit, and was featured in the *Car Design Yearbook 4*. We said back then that the concept's frontal styling was too extreme and that the production XL-7 might need to adopt a more conventional format. Although the side profile is still similar, all the individual design elements are changed, and the boatlike frontal style has indeed been moderated in favour of narrow, angular headlights and an undistinguished horizontal bar grille.

Modern Suzuki styling cues are difficult to spot: the triangular theme to the indicators and bulging wheel arches are perhaps the two principal examples. These apart, only the prominent 'S' on the grille and tailgate save the XL-7 from a complete SUV identity crisis.

Inside, the dashboard and doors feature gently curving block surfaces with black panels carrying instruments and controls. Again, the interior is neat and smooth,but undistinguished in relation to the competition. As evidence of its more carlike crossover status, front-wheel drive is standard, leaving Suzuki's trademark all-wheel drive to the extra-cost options list.

LBD 018

Tata Elegante

Power	149 kW (200 bhp)
Torque	280 Nm (206 lb. ft.)
Gearbox	6-speed manual
Installation	Front-engined/front-wheel drive

Tata, as the best-known industrial group in India's booming economy, has ambitions well beyond the already vast car market on the Subcontinent. As well as being involved in a far-sighted programme to develop an ultra-low-cost car for emerging markets, Tata has deep connections with Italy's Fiat and is open about its intention to bring its products to Europe, which it describes as the most demanding car market in the world. Already, some Tata models are sold in certain European member states.

Tata has taken a stand at the Geneva show for the past ten years, and in 2007 it was the turn of the Elegante concept to take top billing. Based on an all-new platform and claimed to be capable of meeting all European safety and emissions standards, the Elegante is designed as the replacement for the Indica model and is the outcome of a partnership with Fiat Auto that has led to the creation of a joint manufacturing plant in India, with Tata benefiting from access to Fiat's turbo-diesel engines.

Although the Elegante includes all the usual features of a smart mid-priced saloon, such as cruise control, satellite navigation and Bluetooth compatibility, its design lacks some of the subtlety expected in Europe. However, judgments are difficult at this stage, for the concept's heavy, overbodied appearance may be misleading: the long, thin headlamps and the gaping bumper intakes seem to be part of a heavy-handed body kit, and simpler shapes may lie underneath.

Tata has often sought assistance on the shaping of its cars from Italy's premier design houses. Elegante is of course a risky name to choose, and some argue that this is a design that does need a helping hand. Yet, one way or another, Tata is a shrewd company that deserves to be taken seriously wherever it operates.

Toyota Auris

Design	Soichiro Okudaira
Engine	2.2 in-line 4 diesel (1.4 and 2.0 diesel, and 1.4 and 1.6 petrol, also offered)
Power	132 kW (177 bhp) @ 3600 rpm
Torque	400 Nm (295 lb. ft.) @ 2000–2600 rpm
Gearbox	6-speed manual
Installation	Front-engined/front-wheel drive
Front suspension	MacPherson strut
Rear suspension	Torsion beam
Brakes front/rear	Discs/discs
Front tyres	225/45R17
Rear tyres	225/45R17
Length	4220 mm (166.1 in.)
Width	1760 mm (69.3 in.)
Height	1505 mm (59.3 in.)
Wheelbase	2600 mm (102.4 in.)
Track front/rear	1516/1512 mm (59.7/59.5 in.)
0–100 km/h (62 mph)	8.1 sec
Top speed	215 km/h (134 mph)
CO_2 emissions	164 g/km

Auris may be an unfamiliar name, but the look and the brand certainly are not. For this is the European-inspired replacement for the Toyota Corolla, the world's bestselling individual model, and the somewhat bland style is designed to take over seamlessly from where the Corolla left off. That car always came across as stodgy and unexciting in its style; the Auris, finished as a show car in a lurid shade of gold appropriate to its name, is tasked with bringing more spark to Toyota's medium-car line-up.

First impressions, both of the Space Concept show car and of the production model revealed a matter of weeks later, disappointed those who had hoped for a progressive style statement. Instead, the Auris's proportion is very similar to the Volkswagen Golf, albeit with much deeper sides and a slightly higher roof for easy access.

The heavy look of the Corolla is still there, but in the new model's defence, the proportions are very family-friendly, with short overhangs for ease of parking. The wide depression running down through the bonnet and front bumper echoes that on the smaller Yaris. The doors are softly shouldered, the sides capped by pointed lamps front and rear. The blackened upper B-pillars try to disguise its tallness, while forward-leaning C-pillars add a certain sporting touch. The complex rear lamps are very fussy compared with the plain rear-end panel design, but more successful is the large glass roof that allows light to flood in and bathe the dark Germanic-looking interior. Notable in the interior is a bridge linking the dashboard and the centre console, carrying a high-mounted gearlever and handbrake, and with storage space below.

The Auris was designed at ED2, Toyota's European design centre in the South of France, and is built in the UK. It may be practical, well made and reliable, but as a design it is disappointing – and certainly not as successful as the Yaris designed in the same studio.

AURIS

AURIS

Toyota FT-HS

Design	Alex Shen
Engine	3.5 V6 petrol/electric hybrid
Power	298 kW (400 bhp)
Installation	Front-engined/rear-wheel drive
Brakes front/rear	Discs/discs
Front tyres	245/35R21
Rear tyres	285/30R21
Length	4323 mm (170.2 in.)
Width	1859 mm (73.2 in.)
Height	1290 mm (50.8 in.)
Wheelbase	2649 mm (104.3 in.)
Track front/rear	1600/1555 mm (63/61.2 in.)
0–100 km/h (62 mph)	4.2 sec

Presented in the appropriately futuristic guise of a dramatic-looking sports car, Toyota's FT-HS hybrid concept packs both a V6 engine and an electric motor and battery pack. Toyota's belief is that the power-boosting advantages of hybrid drive will soon make supercar performance available at a realistic price and in a realistic, practical format.

According to the designers, inspiration for the shape of the FT-HS came from the look of a downhill speed-skier, hence the slingshot shape that curves around the doors and looks tensioned as if it could be released at any time. There is a sense of leaning forward, both in the sports seats and the door aperture and in the wedged stance of the whole vehicle. The cant rail drops down behind the door, leaving a small air intake to direct air to cool the batteries. The side skirts drop close to the ground, enclosing all but a small part of the wheels' circumference, yet the design language is deliberately kept lightweight and agile through Toyota's so-called 'subtractive mass' technique – the removal of non-essential body volume. This is very clear in the large hollows below the headlights and, especially, the rear lights.

The interior features some glorious sci-fi design forms, almost like a futuristic fighter jet. The steering wheel is a rim, and nothing more. The two-tone metallic colours are contrasted by flashes of bright red to energize the driver, perhaps to signify the FT-HS's dual power source.

Toyota has already achieved high environmental awareness through its success with the hybrid Prius. The hybrid sports car is an obvious next step, adding an impressive electric performance boost at little ecological cost. The fact that this latest design study is shown in white shows Toyota's desire to project a clean image. This is a very exciting development and gives fresh hope to all those sports enthusiasts who thought that environmentally responsible cars had to be dull.

HYBRID
SYNERGY
DRIVE

TOYOTA

Toyota Highlander

Design	Calty Design
Engine	3.5 V6 (3.3 hybrid also offered)
Power	201 kW (270 bhp)
Torque	338 Nm (249 lb. ft.)
Gearbox	5-speed automatic
Installation	Front-engined/four-wheel drive
Brakes front/rear	Discs/discs
Length	4785 mm (188.4 in.)
Width	1910 mm (75.2 in.)
Height	1760 mm (69.3 in.)
Wheelbase	2789 mm (109.8 in.)
Fuel consumption	8 l/100 km (35 mpg)

An altogether more sophisticated Highlander was launched at the Chicago Auto Show in 2007. Compared with the model it replaces, the car uses more complex forms and surface language to make it stand out on the road. For example, the headlamps now have a more purposeful shape to them and are taken round into the side of the vehicle, while the front grille, with its fine-tooth-comb venting, carries a much stronger identity.

This new 2008 Highlander brings feature lines into its flanks to help avoid the heavy, slabby feel of many SUVs. The wheel arches are large and defined and are connected along the sill-level feature line, while from the upper edge of the wheel-arch flares these lines extend front and rear to blend into the profiles of the respective lights. Seen from the side, the mass of the car is placed over the rear wheel, but this is lightened by the slight kick-up on the rear quarter-window, which adds a touch of dynamism.

From the rear the Highlander looks rugged, with its large wheel-arch clearance, lower body guard and the tailgate stepped inside the bumper. The interior is now crisper and with sharper edges, giving it a more contemporary and precise feel. A large screen mounted in the centre of the dashboard sits above a wide centre console with neatly finished switchgear and chrome detailing.

Toyota has three mid-size SUVs in its North American line-up, the others being the 4Runner and the FJ Cruiser. A hybrid version of the new Highlander is also available, using a similar powertrain to the Lexus RX400h. Like its predecessor, this latest Highlander is built on the contemporary Camry platform.

Toyota Hybrid X

Design	Toyota ED2
Brakes front/rear	Discs/discs
Front tyres	225/40R20
Rear tyres	225/40R20
Length	4500 mm (177.2 in.)
Width	1850 mm (72.8 in.)
Height	1440 mm (56.7 in.)
Wheelbase	2800 mm (110.2 in.)

With Toyota having established itself as the undisputed market leader in hybrid technology, its chief designer, Wahei Hirai, now believes that the time is right to develop a distinct visual identity for hybrid vehicles. The current Prius already does this to a certain extent, with its strongly arched roofline. Still, Hirai insists, much more can be done to break with conventional car proportions. Whether you like the Hybrid X or not, it will certainly encourage debate among design professionals and amateur critics alike.

Designed at Toyota's ED2 studio in Nice, France, the Hybrid X is a large, undecorated concept with an arched DLO, smoothly sculpted sides and a heavy, vanlike mass of C-pillar at the rear. Its LED headlamps are set into the base of the windscreen and can display written words, like motorway information boards. The rest of the nose is plain bodywork; no grille is included. When viewed from above, the glass roof can clearly be seen to run right through to the tailgate.

The long doors open invitingly outward from the centre to reveal a cabin where organic flowing surfaces connect the front of the interior to the rear. Sound, smell, sight and touch are all part of the designed interior ambience; many of these parameters can be controlled by the driver. As with current hybrids, a screen displays information to the driver about the energy flow path, fuel consumption and other factors; this is now mounted in the centre of the steering wheel.

Too radical for today's tastes, the Hybrid X is nevertheless a hint at what a next-generation Prius might look like, even though Toyota was giving little away as to the engineering layout under the skin.

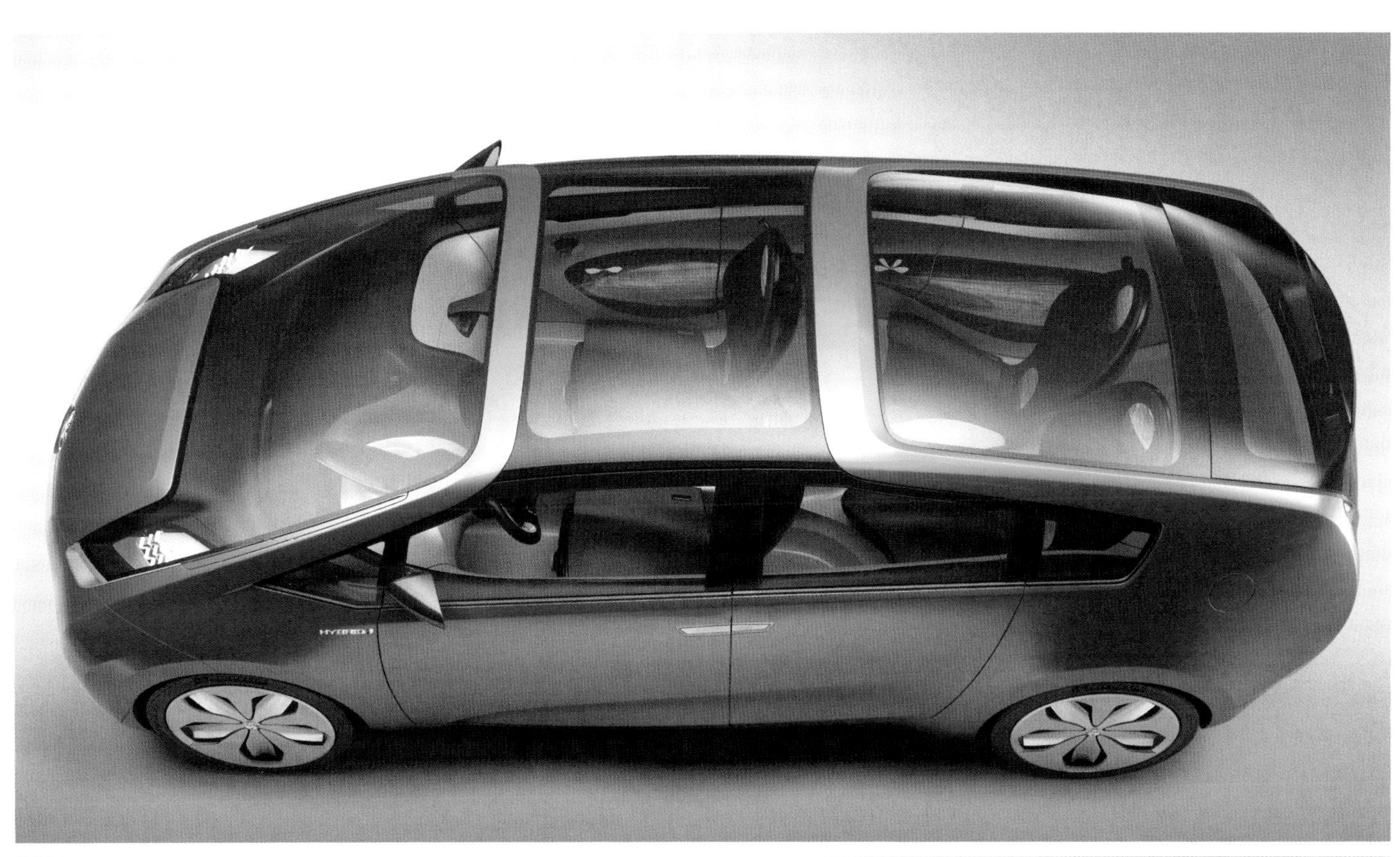

HYBRID

Toyota Tundra

Engine	5.7 V8 (4.0 V6 and 4.7 V8 also offered)
Power	284 kW (381 bhp) @ 5600 rpm
Torque	544 Nm (401 lb. ft.) @ 3600 rpm
Gearbox	6-speed automatic
Installation	Front-engined/four-wheel drive
Front suspension	Double wishbone
Rear suspension	Live axle and leaf springs
Brakes front/rear	Discs/discs
Front tyres	255/70R18
Rear tyres	255/70R18
Length	5329 mm (209.8 in.)
Width	2029 mm (79.9 in.)
Height	1930 mm (76 in.)
Wheelbase	3221 mm (126.8 in.)
Track front/rear	1725/1725 mm (67.9/67.9 in.)
Kerb weight	2227 kg (4910 lb.)
Fuel consumption	14.7 l/100 km (19.2 mpg)

Toyota appointed its US commercial truck operations to design and build the most powerful of the three engines that are fitted to the brand's new Tundra truck, the I-Force 5.7-litre V8. Toyota claims that the Tundra will haul up to 4535 kg (10,000 lb.) in payload, making it a real competitor to the full-size trucks from Ford, GM and Dodge.

The new Tundra is much larger in every dimension than the model it replaces; again, this is a response to taunts in some quarters that the older model, while highly respected for its quality and refinement, was too small to be considered as a full-size truck. The Tundra, on cue, now looks larger and more solid, although it avoids the macho excesses of the bigger US-branded trucks. A wide choice is ensured, with 4x2 and 4x4 drivetrains, three cab styles – Regular Cab, Double Cab and CrewMax – three wheelbases, three bed lengths, three engines and three trim levels.

The broad chrome front-grille surround is the dominant feature emphasizing the power that lies beneath the bonnet. But the design is more about function than anything else. The Tundra is designed as either a work truck, a recreational truck or a luxury family truck. Typical customers include ranchers, foremen and construction workers. Because many of their jobs require the wearing of protective clothing, the interior features large door handles and easy-to-turn control knobs that can be operated even while wearing work gloves. The huge centre console on bucket-seat models is large enough to fit a laptop computer inside or store hanging-file folders – a pickup-truck first.

Features on the new Tundra include a lockable tailgate that has heavy-duty dampers fitted to the hinges so the gate opens smoothly, extra-large side mirrors, a trailer hitch integrated into the frame, a wide-screen reversing camera integrated into the tailgate handle, and an optional premium audio system and Bluetooth phone compatibility.

Tramontana

Design	Josep Rubau
Engine	5.5 V12
Power	537 kW (720 bhp) @ 5750 rpm
Torque	920 Nm (678 lb. ft.) @ 4000 rpm
Gearbox	6-speed manual
Installation	Mid-engined/rear-wheel drive
Front suspension	Double wishbone
Rear suspension	Double wishbone
Brakes front/rear	Discs/discs
Front tyres	245/35R20
Rear tyres	335/30R20
Length	4900 mm (192.9 in.)
Width	2080 mm (81.9 in.)
Height	1300 mm (51.2 in.)
Kerb weight	1250 kg (2756 lb.)

Tramontana is a Spanish company set up by former Volkswagen engineer Josep Rubau. Many automotive industry people dream of designing and manufacturing their own supercar; Rubau has achieved this dream and not only showed a one-off prototype back in 2005, but has also now launched the production version.

Powered by a 5.5-litre V12 engine, the car is clearly inspired by Formula One and uses a carbon-fibre monocoque with some aluminium exterior panels; its two seats are arranged, fighter-jet style, in a tandem layout. The jetlike flip-up cockpit canopy also points to aeronautical inspiration. In fact, an engineer who worked on the Eurofighter jet programme was involved in the car project, too.

Tramontana is focusing its appeal on the fact that the car is handmade: this allows for substantial personalization in return for the hefty – though still not yet officially disclosed – price tag. The company will kit out the interior of a customer's car with special woods and white, gold or leather trim as required. You can be sure of buying a unique piece of Spanish automotive history, and exclusivity is assured as only twelve cars are due to be built each year.

This is clearly a design where form follows function, where the huge mass at the rear is necessary to contain the engine, fuel tank and cooling infrastructure. The body looks somewhat bulky when compared with a Formula One car, yet against the F1 yardstick the wheels look small, especially at the rear. Tramontana might have been better advised to incorporate the wheel arches into the bodywork, but that would then diminish the F1 link. However, although the achievement for a small team to engineer and manufacture such a car is enormous, the real pleasure will be reserved for those brave enough to open their wallets and take the driving seat.

Venturi Eclectic

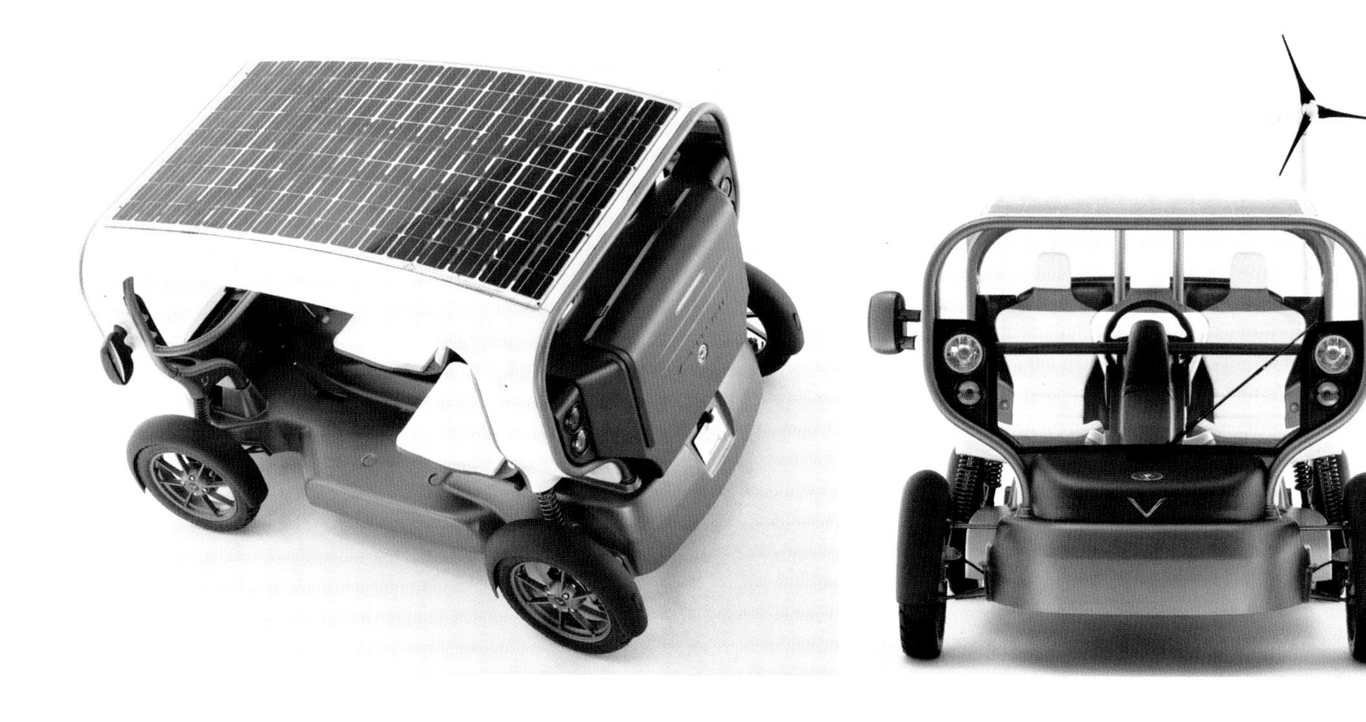

Design	Sacha Lakic
Engine	Electric power with solar assist and wind turbine recharge
Power	16 kW (21 bhp)
Torque	50 Nm (37 lb. ft.)
Front suspension	MacPherson strut
Rear suspension	MacPherson strut
Brakes front/rear	Discs/discs
Front tyres	130/80R17
Rear tyres	130/80R17
Length	2860 mm (112.6 in.)
Width	1850 mm (72.8 in.)
Height	1750 mm (68.9 in.)
Wheelbase	2000 mm (78.7 in.)
Track front/rear	1690/1690 mm (66.5/66.5 in.)
Kerb weight	350 kg (772 lb.)
Top speed	50 km/h (31 mph)

French specialist Venturi, once best known for its sports cars, is now venturing into the area of highly individual and offbeat low-volume models. Yet the Eclectic is a breath of fresh air in more ways than one: not only does it look like a large golf buggy, but also it uses both wind energy and solar power to drive its wheels.

The vehicle makes some other rather grand claims, too, such as being the first autonomous-energy vehicle and the most economical environmentally friendly vehicle ever built. It is certainly the world's first vehicle to use solar and wind power in addition to household electricity for recharging. The nickel hydride batteries will power it at up to 50 km/h (31 mph) for a range of 50 kilometres (31 miles). The extra power generated from the roof-mounted solar panel adds up to 7 kilometres (4 miles) to the range, and an additional 15 kilometres (9 miles) can be captured from wind power if the optional wind turbine is attached while the machine is parked. This may not seem much, but if used only a couple of times a week the Eclectic could easily recharge on sun and wind alone and thus be fully renewably powered.

The overall design and structure take simplicity to a new extreme by having the front and rear almost identical. The roof area is maximized for solar gain, dictating vertical front and rear screens, and the lack of doors and the three-seater configuration make for easy access to any seat. The Eclectic fits the requirements of a quad-class vehicle rather than that of a conventional car, meaning that there is no need for windows or seatbelts.

Venturi expects the Eclectic to appeal to users away from the public highway, such as in the grounds of large hotels, at golf courses and in conference centres. Conceptually it is one of the most interesting cars featured in this book, and a very limited series of hand-built examples has already been produced.

Volkswagen Iroc

Design	Robert Lesnik
Engine	In-line 4
Power	157 kW (210 bhp)
Installation	Front-engined/front-wheel drive
Brakes front/rear	Discs/discs
Length	4240 mm (166.9 in.)
Width	1800 mm (70.9 in.)
Height	1400 mm (55.1 in.)
Wheelbase	2680 mm (105.5 in.)

The fact that the name Iroc is lifted from the word 'scirocco' gives a strong clue to the character and role of this important design study from Volkswagen. After years of worthy but dull models the German company is struggling to create excitement around its products, and the original Giugiaro-designed Scirocco of 1974 was a smart-looking, fun-driving coupé that is remembered with particular affection.

Finished in the same lurid Viper Green as the original Scirocco show car, the Iroc has a similarly low stance and also echoes the original in its rising window line. But that is about as far as the parallels go, for the Iroc shows off the new Volkswagen face for sports models with its deep and wide six-sided honeycomb grille, the border of which forms a bumper-like structure. The xenon headlamps and flat bonnet have echoes of the current Golf, as does the A-pillar position and the visual strength in the wide C-pillar as it tucks round to meet the rear window. The long rear side window rising up gives the Iroc an angular poise on the road, while the long, low roof finishes off with a chunky spoiler that shields the rear screen. The growing shoulder line flows into the rear lamps where it blends into a rounded tailgate with simple horizontal features.

Black carbon guards run around the base of the car and give a forward-pointing stance when the vehicle is viewed from the side, while at the rear the bumper sits high, leaving the tailgate shallow. Stepped-out wheel arches pushed right to the corners at the rear serve visually to widen the rear aspect.

Green detailing and trapezoidal shapes are carried through to the interior to create a poorly resolved and uncomfortable mix of themes; any production interior would have to be much better executed to do justice to this promising design.

Volkswagen Neeza

While Auto China 2006 in Beijing hosted a wealth of design studies, only the Volkswagen Neeza came from a non-Chinese brand. Yet, significantly, it was not only aimed at the Chinese market but also designed in China. This study, says Volkswagen, opens a window on the future of new models from Shanghai Volkswagen, one of the very first Western automakers to enter the Chinese market. It is claimed the Neeza – named after a Chinese folk-tale hero – represents complete harmony between German technology and Chinese culture.

The Neeza is a medium–large station-wagon built on a stretched version of the Golf platform, though there is little sign in its overall shaping of the coupé crossover influence cited by VW in its specification. And aside from what are described as traditional Chinese engravings in the frontal grille, it is a serious stretch of the imagination to find the traces of the Chinese culture claimed by VW.

What is most evident from a design perspective is how VW's Shanghai-based designers have tried to lengthen the car both physically and visually. Front and rear overhangs are very long, creating an awkward overbodied look, and the long, sloping roof, the straight feature line running through the doors, and the grey plastic skirt around the lower body all contrive visually to stretch the car. From the rear the Neeza looks uncannily similar to Audi's prestigious Q7.

The interior is largely conceptual in beiges and bright reds: long elliptical shapes try to reflect the streamlined nature of the exterior, and the full-length glass roof is said by VW to express the philosophy of open spaces. Mood lighting is able to turn the cabin a variety of shades, and VW has intentionally omitted all electronic equipment so that the user can install the very latest devices.

Neeza

Volkswagen Tiguan

Design	Klaus Bischoff
Installation	Front-engined/all-wheel drive
Brakes front/rear	Discs/discs
Length	4400 mm (173.2 in.)
Width	1850 mm (72.8 in.)
Height	1690 mm (66.5 in.)

Designed at Volkswagen's headquarters in Wolfsburg, the Tiguan is a more mainstream development of the coupé-like Concept A crossover featured in the *Car Design Yearbook 5*: at the time the Concept A was billed as a mini-Touareg, and the appearance of the Tiguan (a name chosen in a German magazine's reader poll) confirms that this will indeed be VW's middleweight SUV to take on the likes of the Toyota RAV4 and the Land Rover Freelander.

The latest concept is very close to the eventual production version, which is due to reach the showrooms by the end of 2007. The front, with its U-section grille frame now dark-anodized rather than bright aluminium, gives a similar muscular impression to the Concept A, but to the rear of the A-pillars the Tiguan is totally different. Gone are the racy low-rider coupé look and the sloping tailgate; in its new identity, the vehicle is an upright, secure and safe-looking 4x4 with no ambiguity or concessions to fashion.

Overall, the Tiguan is well resolved, with neat production-ready detailing and an absence of gimmicks that would be difficult or expensive to incorporate in volume manufacture. It communicates a good sense of power and off-road ability thanks to its tough-looking front end; the piercing headlamps, forward-pointing stance and large squared-off wheel arches are particularly effective. The dark skirt trim around the car adds to its off-road look and reduces its visual weight.

A concept feature unlikely to make it to production is the contrast colour built into the treads of the tyres, though VW says the anthracite-coloured alloy wheels will make it through to the showroom. Inside, however, there is little innovation, with the design and layout being carried over from the Golf Plus.

TIGUAN

Volvo C30

Design	Simon Lamarre
Engine	2.5 in-line 5 turbo (2.4 in-line 5, 1.6, 1.8 and 2.0 in-line 4, 1.6 and 1.8 in-line 4 diesel and 2.4 in-line 5 diesel, also offered)
Power	220 kW (295 bhp) @ 5000 rpm
Torque	320 Nm (236 lb. ft.) @ 1500–4800 rpm
Gearbox	6-speed manual
Installation	Front-engined/front-wheel drive
Front suspension	MacPherson strut
Rear suspension	Multi-link
Brakes front/rear	Discs/discs
Length	4252 mm (167.4 in.)
Width	1782 mm (70.2 in.)
Height	1447 mm (57 in.)
Wheelbase	2640 mm (103.9 in.)
Track front/rear	1535/1531 mm (60.4/60.3 in.)
Kerb weight	1350 kg (2976 lb.)
0–100 km/h (62 mph)	6.7 sec
Top speed	240 km/h (150 mph)
Fuel consumption	8.7 l/100 km (32.5 mpg)
CO_2 emissions	208 g/km

The new compact C30 is part of a fresh and lucrative market trend: the emergence of the premium-branded small car. Examples are the Alfa 147, the Mini and the BMW 1 Series. For Volvo the newcomer has a very specific role: as a stylish and dynamic model that is also affordable, it is charged with drawing young people with accelerating careers into the Volvo brand, where it is hoped that they will continue buying a succession of bigger and more expensive Volvos as their salaries rise.

The C30 comes with Volvo's signature front-end design and its low, wide grille, yet the side view really differentiates this car not only from the rest of the Volvo range but also from any potential competitor car. Nothing has a profile even remotely similar. It is a distinctly sporty model characterized by the forward-sloping dark-glass rear hatch and downward-sloping roofline. Obvious parallels can be drawn with the Volvo 480ES from the 1980s and even the P1800ES in the early 1970s, but this is probably of little consequence for today's buyer.

Mouldings around the wheel arches emphasize a certain sportiness, as does the bold shoulder line that runs right through from front to back giving the C30 a sense of strength, which of course it has because it is a Volvo.

Inside, personalization is a strong theme. Buyers can opt for a variety of finishes to the signature floating centre stack and can even select sporty red carpets if so desired; otherwise a mix of dark grey and black trim and aluminium features heavily.

Designed for people who have a fast-moving lifestyle, the C30 is the perfect entry-level model to draw new customers into the steadily expanding Volvo mainstream range. It is safe, stylish, well thought out, and just about as practical as a small coupé-like four-seat hatch could hope to be.

NEW YORK
VOLC30

Volvo V70

Design	Steve Mattin
Engine	3.0 in-line 6 (2.5 in-line 5, 3.2 in-line 6, and 2.4 in-line 5 diesel, also offered)
Power	210 kW (282 bhp) @ 5600 rpm
Torque	400 Nm (295 lb. ft.) @ 1500–4800 rpm
Gearbox	6-speed manual
Installation	Front-engined/front-wheel drive
Front suspension	MacPherson strut
Rear suspension	Multi-link
Brakes front/rear	Discs/discs
Length	4823 mm (189.9 in.)
Width	1861 mm (73.3 in.)
Height	1547 mm (60.9 in.)
Wheelbase	2816 mm (110.9 in.)
Track front/rear	1588/1586 mm (62.5/62.4 in.)
Kerb weight	2350 kg (5181 lb.)
0–100 km/h (62 mph)	7.2 sec
Top speed	245 km/h (152 mph)
Fuel consumption	11.2 l/100 km (25 mpg)
CO_2 emissions	267 g/km

It goes without saying that as the latest car from Volvo – and a station-wagon, too – the new V70 promises to be safer than ever, and perhaps even a candidate for the safest car in the world. That is what we have come to expect from Volvo; but what about the rest of the design?

The car is now built on the S80 platform, so it is larger than the older model and promises better driving characteristics. Many structural elements are common to the S80, in fact, including some of the interior architecture. Two versions were launched at the Geneva show in 2007, the standard V70 station-wagon and the crossover XC-70 (above right and opposite, top), the latter being essentially a V70 with four-wheel drive, higher-riding suspension, a more rugged off-road look and the new style of front grille.

When viewed from the front, the new V70 looks pretty similar to the outgoing model. From the side, a slightly narrower shoulder can be seen, and the concave sides have gone; it is at the rear that most of the change is noticeable. Gone is the flat tailgate: now, the number-plate is recessed and the rear bumper projects strongly. The vertical tail lights have become much bigger, to the extent that most of the lights now rise with the tailgate. The result is some awkward shut lines and a more cluttered look, though this design does suit the XC-70 version better, where the rear bumper is black and an aluminium stone-guard is fitted underneath.

The windscreen is now more steeply raked, and blackened upper pillars comply neatly with the latest fashions. Innovations inside include the world's first integrated dual-stage children's booster seat. Volvo's options include a US-friendly powered tailgate, laminated glass windows, a sliding load-floor, active bi-xenon headlamps and adaptive cruise control with radar collision warning.

VOLVO
XZX 240
V70

Volvo XC60

Design	Steve Mattin
Engine	3.2 in-line 6 bioethanol
Power	198 kW (251 bhp)
Torque	340 Nm (461 lb. ft.)
Installation	Front-engined/all-wheel drive
Brakes front/rear	Discs/discs
Length	4560 mm (179.5 in.)
Width	1915 mm (75.4 in.)
Height	1635 mm (64.4 in.)
Wheelbase	2770 mm (109 in.)
0–100 km/h (62 mph)	8.5 sec
Top speed	230 km/h (143 mph)
Fuel consumption	12.3 l/100 km (23 mpg)

The XC60 seems an obvious place to begin expanding the Volvo range. The larger XC90 has been well received and is selling strongly, and BMW's success with both the X3 and the X5 is clear evidence that a pair of 4x4 models can sit side by side in a premium model range.

Volvo has tried to make the car visually softer and more friendly than typical larger SUVs, as befits its role as a crossover rather than a pure SUV; it is also a (small) nod to the growing public and official backlash against gas-guzzling 4x4s. This concept is a deliberate pointer by Steve Mattin, Volvo's design director, to the new and more emotional design language being explored by Volvo. The surfaces on the bumpers and doors are more sculpted than on current production models. The stance is distinctly high-tail, the overall look more sporting and more BMW-like; this marks a departure from the understated quietness of Volvo design today. Nevertheless, the strong Volvo shoulder is, as ever, present and is clearly visible at the rear, where it is incorporated into the very distinctive rear lamp design.

Inside, the XC60 is almost entirely white: the effect is somewhat technical and cold and is at odds with the warmer-looking exterior. Mattin describes the broad floating centre console – now a Volvo characteristic – as like a giant iPod.

A new technical feature shown on the XC60 is Volvo's City Safety system. This uses short-range radar to apply the brakes in slow-moving traffic if it senses the risk of an accident. Many of the technical features and design cues of the XC60 also appear on the new V70 estate, announced just a month later.

The XC60 is a great-looking car that comes across as dynamic and well resolved. Volvo says it will enter quantity production in 2009, but for many impatient customers that is not nearly soon enough.

Zagato Diatto Otto Vu

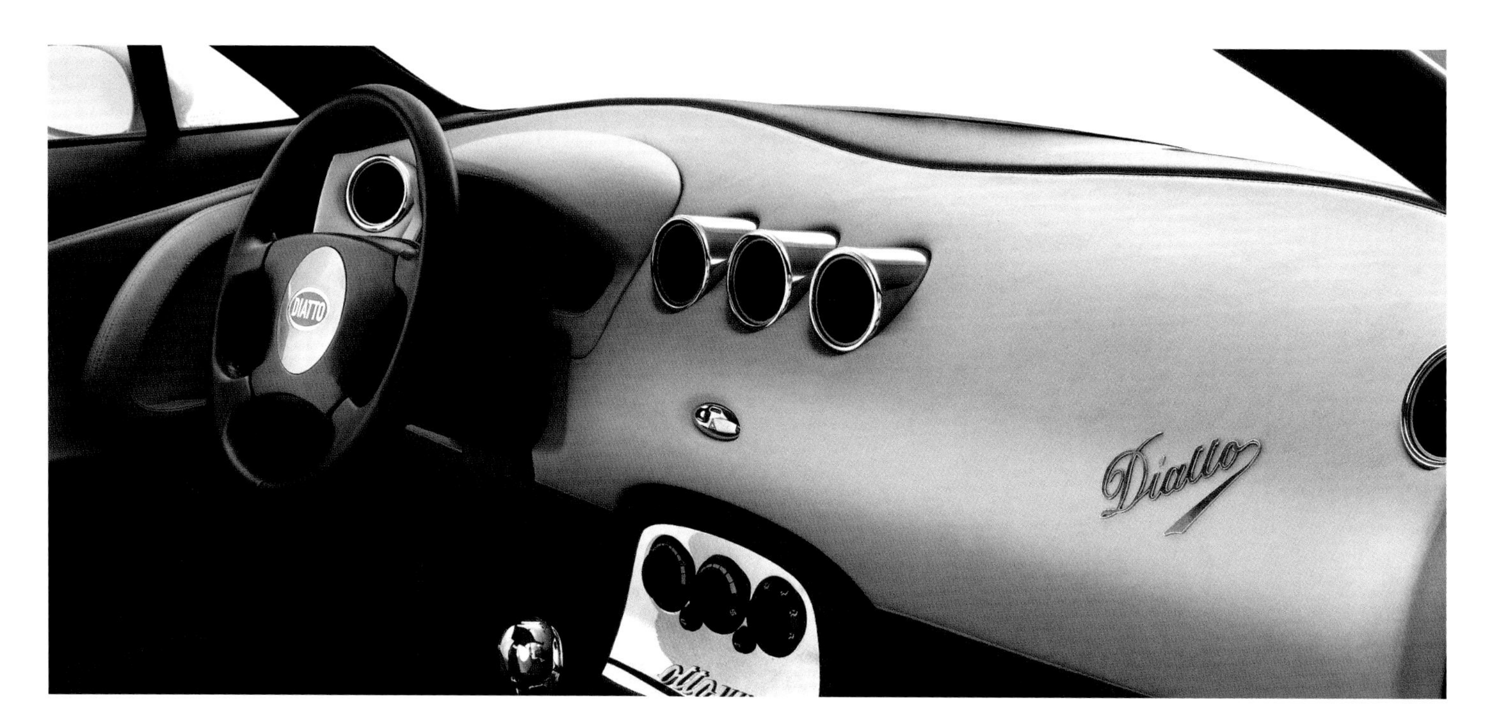

Design	Zagato
Engine	4.6 V8
Power	373 kW (500 bhp)
Installation	Front-engined/rear-wheel drive
Brakes front/rear	Discs/discs

With the intention of celebrating the hundredth anniversary of the Diatto marque, two wealthy collectors formed a partnership with Italian coachbuilder Zagato to help breathe new life into the dormant name. Diatto's relationship with Zagato – long known as one of Italy's most eccentric designers – dates back to the 1920s, adding a further historical element.

Nevertheless, the resultant design is thoroughly modern. The Otto Vu has similar proportions to the current Maserati Coupé, but is even more voluptuous. The model was shown with no bonnet or boot shut lines so the surface form can be appreciated in full. The front end dips towards the chrome oval grille, which is flanked by two large air intakes. Two sharp crease lines rise up along the fenders visually to widen the front of the car. From the side the fuller rear fender has a bluff rounded shape that lightens the rear end, making the vehicle look shorter and more nimble. The two crease lines that run just above the wheels add to the sense of Italian flair as well as complementing the sporting nature of the design. Rounded features are commonplace: the rear quarter-windows, the door mirrors, the rear screen and lamps, are all curved in shape and point to the heritage of the brand.

The very simple interior marries a pale cream dashboard with polished metal air-vent bezels, a small conventional instrument display in front of the driver, and a bright metal surround to the gear lever. Interestingly, the Diatto logo, with its white lettering on a red oval bordered with a string of red dots, is almost identical to that of Bugatti.

This is an attractive car born out of a passion for Diatto, and an interesting mix of traditional body-surfacing techniques on a modern-day platform, thought to be that of the Maserati Coupé.

ZAGATO

ZAGATO
milano

LAND ROVER
CONTINENTAL

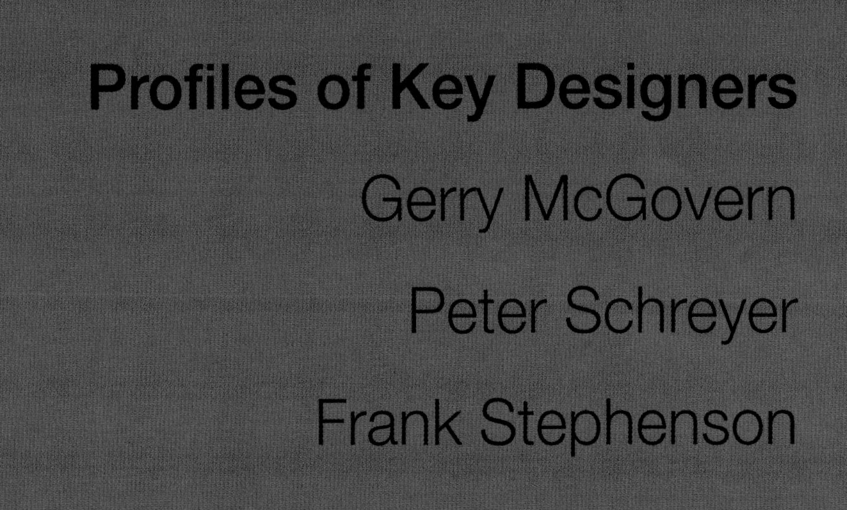

Profiles of Key Designers

Gerry McGovern

Peter Schreyer

Frank Stephenson

Gerry McGovern

Gerry McGovern officially took full charge of all Land Rover design in summer 2006, on the retirement of Geoff Upex, but his association with the iconic four-wheel-drive marque goes back much further. He first came to widespread public attention in a Land Rover context in 1997, with the launch of the first-generation Freelander, and after a long spell in other Ford-group design roles, he returned to Land Rover as director of advanced design in April 2004.

To sports-car enthusiasts, however, McGovern is best known as the man who penned the 1995 MGF, the mid-engined two-seater sports car that put MG back on the international map as a builder of stylish, fun-to-drive and affordably priced roadsters. For many years the MGF, and its mildly reskinned successor, the TF, were top sellers in Europe, and the design is set to live another day as China's Nanjing automaker, which bought the collapsed MG Rover organization in July 2005, puts the model back into production in the UK and the US as well as in China itself.

McGovern was born in Coventry, the cradle of UK carmaking, in 1956, and describes himself as having been a keen artist at school before his interests shifted towards industrial design. His art teacher introduced him to the design director of Chrysler, which had recently taken over such well-established European marques as Hillman, Sunbeam and Simca, and which had its major European design facility at Whitley, near Coventry.

McGovern was now set on a career in car design, his personal inspiration being the Lamborghini Miura. Chrysler's highly respected design boss, Roy Axe, was to have a major influence on McGovern's career, as on the many other young designers he nurtured. Chrysler sponsored the young McGovern through Lanchester Polytechnic (now Coventry University), where he did a degree in industrial design, and from there he went on to the Royal College of Art in London, specializing in automotive design.

In 1978, big business deals intervened. McGovern was doing a short stint for Chrysler in

Top
The rights to the MGF design, updated in 2002 as the MG TF, were acquired by China's Nanjing in 2005, and UK production is set to resume.

Opposite top
In 1985 the MG EX-E was seen as a sign that MG was set on the path to making high-performance cars.

Opposite bottom
With the Continental concept, Ford tried to re-establish Lincoln as the American luxury-vehicle brand.

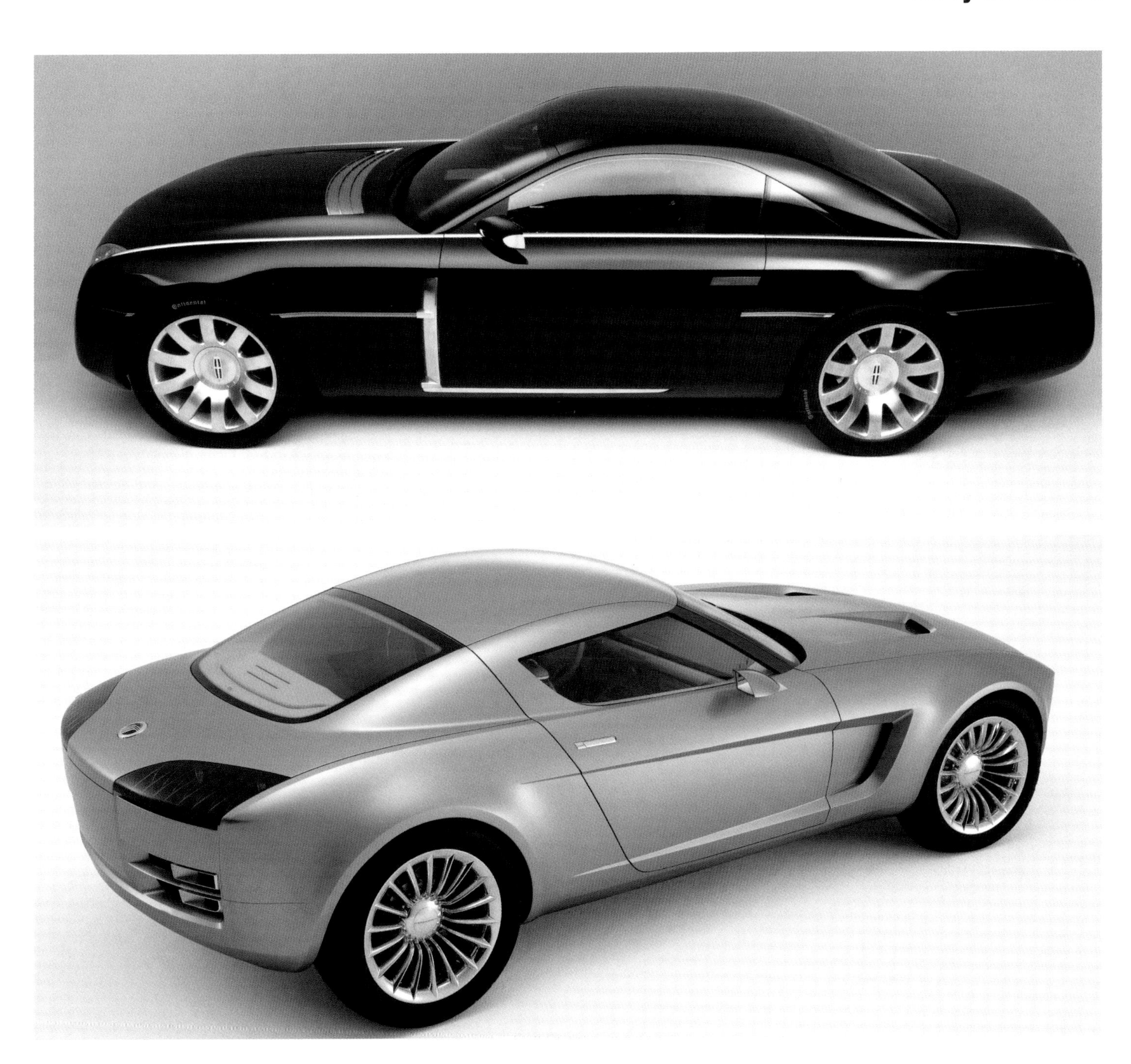

Opposite
Lincoln's Navicross continued the themes developed on the Mk9 concept, but added the more distinctive waterfall grille incorporating the headlamps.

Top
Contemporary design language cemented a powerful image for the gloss-black Lincoln Mk9.

Above
The Mercury Messenger concept combined a number of striking features, including a wraparound windscreen, short overhangs and a vertical rear end.

Detroit just as Chrysler sold its UK facility to Peugeot. He chose to return to the UK as a senior designer for Peugeot; it was there that he worked alongside other designers, including Peter Horbury and Moray Callum, who were later to become key industry figures.

Roy Axe became design director of the Rover group in 1982, and lost little time enticing McGovern to join him. One of McGovern's first cars was the exterior of the very well-received 1985 MG EX-E show car, which predated the production MGF, the first car on which he had overall design responsibility. His first actual production-car responsibility was the exterior design for the coupé version of the Rover 200.

In 1999, a year before Ford bought Land Rover from BMW, Ford's US arm hired McGovern to revive design at Lincoln-Mercury; Lincoln had

MORETTI
GRUNDIG
TRIBERTI
MIURA—S

not had a dedicated design director for more than twenty years. McGovern recruited key new design talent, and, with a series of progressive concepts, challenged popular notions of Lincoln design and showed a possible route forward for the neglected and slow-selling premium brand. Yet McGovern's bold approach was not adopted by Ford management, possibly because it was too closely associated with the regime of CEO Jacques Nasser, who had left the company after rising losses and boardroom disagreements.

McGovern returned to the UK in 2003 as creative director of a venture that seemed highly innovative at the time: Ingeni, Ford's design and creativity centre, in Soho, London. Sadly, with the Ford group haemorrhaging cash in too many directions, Ingeni all too soon had to close its doors, and McGovern rejoined Land Rover, as director of advanced design, in 2004. Just over two years later, he took over full responsibility for all Land Rover design.

Yet McGovern still cites the first Freelander as his most challenging design task: 'It was taking Land Rover into a market it had not been in before, and a lot of people were not convinced we could do it', he told the *Car Design Yearbook*.

And, on the evidence of the momentum provided by that first Freelander, it is clear that Land Rover design is in the best possible hands. The Freelander kicked off a worldwide trend for smart but capable compact SUVs, made the Land Rover name accessible to hundreds of thousands of new customers worldwide, and established the template for the succession of record-breaking years that Land Rover has been enjoying since it became part of the Ford group.

Opposite
The 1960s Lamborghini Miura-S is the car that has inspired Gerry McGovern throughout his career.

Above
The original Freelander (front two vehicles) was not only a great design, but also became the defining product in opening up a new market segment for SUVs that were light, easy and pleasant to drive on-road.

Peter Schreyer

It is a measure of Peter Schreyer's standing in the world of international automotive design that eyebrows were raised when, in September 2006, Kia Motors of Korea announced that he was to join the company as its head of global design. Schreyer, who had a hand in several pivotal designs that went on to become automotive icons – the new VW Beetle and the Audi TT are examples – is considered quite a catch, and by hiring a designer of international repute at the top of his game, Kia has signalled that it is deadly serious about building cars with state-of-the-art quality and style.

'Mr Schreyer's appointment will add to Kia's significantly increasing emphasis on automobile design and support the ongoing regeneration of its model range', announced the company. 'He will be responsible for implementing the company's design vision in line with its brand values that target a young-at-heart and adventurous target customer base.'

Schreyer's move to Kia came after a successful stint at Volkswagen as head of advanced design. He had joined Volkswagen only in 2002, following a shake-up of the Audi design department that saw him replaced by Walter de' Silva. Schreyer had been chief designer at the premium brand for eight years, during which time Audi's standing on the world stage grew enormously thanks to rock-solid engineering and the trend-setting design that went with it.

Despite a flourish of influential concepts and highly successful passenger cars at Volkswagen, such as the Golf, the Passat, the Eos and the new Beetle, Schreyer missed out on becoming head of Volkswagen's newly established advanced design studio in Berlin, the job going to Skoda design boss Thomas Ingenlath. Schreyer then played a relatively low-key role in Volkswagen design circles.

Among other designs Schreyer is credited with at Volkswagen are the spectacular Concept R roadster – the mid-engined two-seater developed under the Volkswagen group's then head, Bernd Pischetsreider – and the novel GX3

Top
Few cars go down as genuine design classics, but the 1998 Audi TT will certainly be one. The simplicity and originality of the design mark it out for admiration that spreads well beyond the automobile community.

Opposite
Like the Audi TT, the new Beetle shows a symmetry in side view, in this instance an overall archlike profile.

WOB AW 975

three-wheeled fun car, which was seriously considered for low-volume production.

But it was his years at Audi that made Schreyer truly well known. During his eight-year stint at the premium carmaker based at Ingolstadt, Bavaria, he developed the classic Bauhaus-inspired design language that is now reflected across the complete Audi line-up. The most notable of these achievements is the TT, in the design of which Schreyer played a major role, and which went on to acquire cult status well beyond the design community. Other highlights include the highly regarded 1997 Audi A6 sedan, the adventurous aluminium-bodied A2 supermini and the well-received A8 luxury sedan, the model that put Audi on the map as a credible rival to Mercedes-Benz and BMW.

Schreyer is also credited with significant input into that other cult car, the reborn VW Beetle. This, he says, was one of his toughest design challenges, the struggle being to keep management on board and convince them of the design processes involved.

Born in 1953, Schreyer began his design training at the Fachhochschule in Munich in 1975, graduating to the world-renowned Royal College of Art in London four years later. Looking back, he says it was Bertone's Lancia Stratos,

Above
Schreyer's Volkswagen Concept R was VW's first exploration of design themes for a mid-engined, two-seater sports roadster. The design's multiple curves tightly wrap the occupants in a protective motion, echoing the way the bodywork wraps around the wheels.

Opposite
The Volkswagen GX3 was a design exercise in the unconventional and the more extreme forms of motoring, adopting a three-wheeler layout for maximum driving sensation.

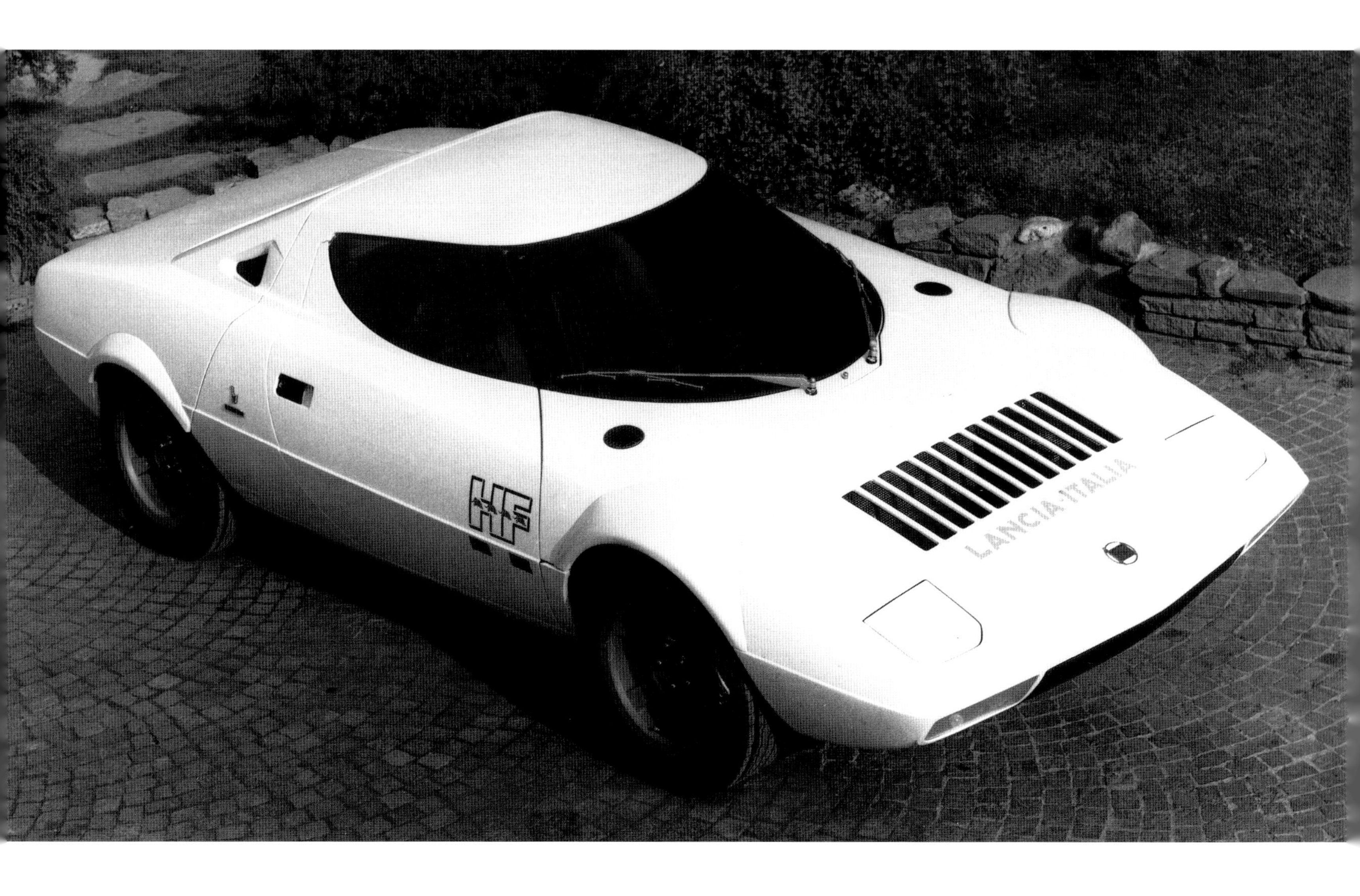

the elegant but devastatingly effective rally car of the 1970s, that inspired him to become a car designer, while among his inspirations from outside the automotive world he cites individuals as diverse as Frank Zappa, Salvador Dalí, Miles Davis and his grandfather.

From the RCA in London, Schreyer went to work at Audi as an exterior designer; his responsibilities soon broadened, and his first major input into a production vehicle was the interior of the then-new Audi 80.

Schreyer is a lover of fine art and has a healthy respect for the work of his fellow automotive designers. Among the designs he picks out as particular favourites is the 1991 Audi Avus concept by J. Mays, Martin Smith and Graham Thorpe.

Schreyer's appointment at Kia is, on his own admission, a rare opportunity to help to define the identity of a whole brand. 'It's a blank sheet of paper', he says. 'There is no history, no heritage, so we will have to begin creating one.'

As evidenced by its major investments in European manufacturing, engine-building, research and development, and design, Kia is gearing up for a major push into Western markets – the very markets in which customers are the most fanatical about the way their vehicles look, the way they feel inside and out, and the way they drive.

Designers at Kia's four international design centres – Tokyo, Frankfurt, Los Angeles and Namyang in Korea – have been responsible for an impressive output of motor-show concept cars over the last few years. This is clear evidence of the design talent that is eager to surface, and is also proof that Kia has for some while been seeking a unified direction for its design. Schreyer's arrival is sure to provide a

much clearer focus to the company's worldwide design work: Schreyer's central task, says the company, will be to mould the future face of the brand and models from the same clay.

To date, and with very few exceptions, Kia's production vehicles have erred on the side of conservatism rather than progressiveness in their exterior signatures. Steeped in the ethos of Audi, where purity of design is paramount and attention to detail is legendary, Schreyer is likely to imprint a much more solid and systematic template on to Kia's design. Greater design flair, yet also a more consistent set of cues across the entire line-up, are likely to be the first fruits of his design regime, although few are yet prepared to give any clues as to the strategy that is likely to be adopted.

As Kia continues to develop models and expand its presence across all key regional markets, it has chosen a design leader who can enhance the company's range of vehicles by accurately reflecting modern aesthetics and regional and cultural sensibilities. For Schreyer, this is a once-in-a-lifetime chance to build the identity of a worldwide brand; the most recent comparable opportunity was when Volkswagen bought the very much down-at-heel Skoda brand and, through consistent design and quality engineering, turned it into a respected and distinctive global player able to compete with world-class brands in terms of customer visibility and appeal.

It will be Schreyer's task to create the clear identity that will allow Kia to achieve a similar scale of transformation – though, given the fact that Kia's range already extends from the budget Picanto hatchback via an assortment of SUVs and crossovers to the executive-sized Opirus sedan, the challenge he faces is greater still.

Opposite
Bertone's creation for Lancia, the Stratos, was the inspiration behind Peter Schreyer's decision to become a car designer.

Above
Audi's 1997 A6 is still seen as a quietly understated yet highly sophisticated product and has been enjoyed by millions of owners.

Frank Stephenson

He may be best known for his work on the iconic new-generation Mini, but Frank Stephenson has a wealth of experience in automotive design that stretches, quite literally, from some of the world's smallest cars to the fastest and most glamorous, in the shape of Ferrari and Maserati.

But even greater heights could be still to come. Now head of vehicle design at Fiat, Stephenson has helped to bring about an important design and commercial renaissance at the once-struggling Italian volume carmaker. The recent Grande Punto and this year's new Bravo have been much lauded for their stylishness, the Punto proving such a hit that it has propelled Fiat back up the small-car sales charts to its customary top-five position. Yet the really big one will be a much smaller car: the reincarnated Fiat 500.

Just like the Mini, the baby Fiat dates back to the 1950s and is a tiny car with a huge amount of sentiment and emotion wrapped up in it. And, also just as with the Mini, a new design is being developed that seeks to tap into this deep reservoir of affection, translating the appeal of the original into a modern context and, most importantly of all, maintaining faith with the defining spirit of the original.

Prior to his arrival at the mainstream Fiat brand in 2005, Stephenson had been with ultra-glamorous Ferrari and Maserati, also members of the Fiat group. Yet his background provided little clue to the career he was to take up. He was born in Casablanca, Morocco, to a Norwegian father and a Spanish mother – coincidentally, on the exact same October day in 1959 that the original Mini was launched at the London Motor Show. Until he was seven, he spoke only French and Arabic; at eleven, inspired by a glimpse of a Ferrari Dino 246, he began to draw cars – a passion that would eventually lead him to train as a car designer at the Art Center College of Design in Pasadena, California, after attending high school in Madrid, Spain.

Stephenson's first job in car design took him back across the Atlantic to the Ford design studios in Cologne, Germany. Soon he achieved fame – or notoriety – as the man who designed

Top
The reborn Mini, designed by Frank Stephenson, has achieved remarkable success.

Opposite top
It was the Ferrari Dino 246 that first made Stephenson want to draw cars.

Opposite bottom
Pininfarina's design of the Ferrari F430 was overseen by Frank Stephenson. The front of the car references the 'shark nose' from Ferrari's 1960s racing cars.

PIRELLI P ZERO

Above
Stephenson rates the MC12 as one of his greatest challenges: the task of combining Maserati design aesthetics with highly sophisticated race-car aerodynamics called for great sensitivity and skill.

Opposite
Few would question the beauty of the Ferrari 612 Scaglietti, the first design Stephenson led after joining Ferrari in 2002.

the very large rear spoiler fitted to the dramatically fast Sierra Cosworth.

Stephenson's eleven years at BMW, from 1991, began with a role as a senior designer in the Munich studio. He worked on a variety of projects, including the 1999 BMW X5, the first 4x4 to break away from the boxy look and agricultural driving characteristics for which SUVs had until then been notorious. Only recently – in 2006 – has this model been replaced, and it is telling that, while the new version is larger, it retains all the design cues of the original. Yet Stephenson is most famous for penning the shape of the reborn BMW Mini Cooper, launched in 2001. Many designers contributed to that programme, but it was Stephenson's acute sense of faithfulness to the original that helped make the project successful beyond even BMW's wildest dreams.

Stephenson moved to Ferrari in 2002, and was quickly able to put his stamp on showroom models. The 2005 Coupé GranSport was the first Maserati project on which his fingerprints are evident, while on the Ferrari side of the business he was able to input some significant late changes into the new 612 Scaglietti of 2003. Such models as the Maserati MC12 and the all-new Ferrari F430, both of 2004, have met with great acclaim, and Stephenson himself cites the MC12 as one of his best designs, even though it was a major challenge to combine attractive aesthetics with race-car aerodynamics.

As with much design in this modern era, projects generally tend not to be the work of a single individual but, rather, the combined efforts of a large team, overseen by a design manager. This is especially true within the Fiat empire, owing to its long-standing tradition of using the expertise of the famous Italian design houses. Such programmes require especially sensitive supervision, and an example of this is the success with which Stephenson oversaw the

Above
Maserati's latest coupé, the GranTurismo, shows the direction for Maseratis of the future, set to challenge the very well-received new models from a fast-growing Aston Martin.

Opposite
Maserati's Quattroporte, although designed by Pininfarina, was overseen by Stephenson. The high-performance luxury sedan has proved to be Maserati's first real success in the four-door segment, following many years of products that missed the mark in terms of style and quality.

introduction in 2003 of the Maserati Quattroporte and the Ferrari 612 Scaglietti, both designed by Pininfarina.

Perhaps the real strength that Stephenson brings to his work is not only his capacity to be a great designer in his own right, but also his ability to understand the importance of what went before, and how to nurture historic influences as an inspiration for new contemporary designs. He did this with the Mini, the remarkable success of which astonished almost everyone in the car business. This was (and is) an ongoing job at Ferrari, too, where the desire to push advanced design to its limits has to be tempered with reverence for the marque's uniquely glorious back catalogue.

And now at Fiat Stephenson's skills are being put to a very public test as the new Fiat 500 competes for the hearts not only of all those who remember the original with such affection but also of younger buyers, for whom the modern reincarnation of the classic must be fun to drive, cool to be seen in, and, above all, relevant to today's image- and value-conscious buyer.

If Stephenson can re-create the sense of fun and style that is felt when driving a Mini, albeit in a package that is cheaper and with less scope for costly designers' whims, then he will have cleared the first major hurdle in inspiring a new golden era in Fiat design. All he will then need to do is to cultivate the right creative climate to see a steady output of inspired shapes that will maintain the momentum of the design renaissance under way and carry the brand into the next decade profitably and in style. And that's where the real long-term vision comes in.

Technical Glossary

Where the New Models Were Launched

Major International Motor Shows 2007–2008

Marques and Their Parent Companies

Technical Glossary

Specification tables

The following list explains the terminology used in the specification tables that accompany the model descriptions. The amount of data available for any given model depends on its status as a concept or a production car. More information is usually available for models currently in or nearing production.

engine
Engine size is quoted in litres, and refers to the swept volume of the cylinders per crankshaft rotation; 6.0, for example, means a 6 litre (or 6000 cc) engine. 'In-line' or 'V' followed by a number refers to the engine's number of cylinders. An in-line 4 engine has four cylinders in a single row, while a V8 engine has eight cylinders arranged in a V-formation. A flat-four engine has four cylinders lying in a horizontal plane, two opposing each other. In-line engines of more than six cylinders are rare today because they take up too much packaging space – an in-line 12, for instance, would require a very long bonnet. Only Volkswagen makes a W12, an engine with its twelve cylinders arranged in a W-formation. The configuration of cylinders is usually chosen on cost grounds: the higher the car's retail price, the more cylinders product planners can include.

power
Engine power is given in both metric kilowatts (kW) and imperial brake horsepower (bhp). Both are calculated at optimum engine crankshaft speed, given in revolutions per minute (rpm) by manufacturers as a 'net' measurement – in other words, an engine's output after power has been sapped by other equipment and the exhaust system – and measured by a 'brake' applied to the drive shaft.

torque
Simply the motion of twisting or turning, in car terms torque means pulling power, generated by twisting force from the engine crankshaft. It is given in newton metres (Nm) and pounds feet (lb. ft.). The higher the torque, the more force the engine can apply to the driven wheels.

gearbox
The mechanical means by which power is transmitted from the engine to the driven wheels. There is a wide variety of manual (with a clutch) and automatic (clutchless) versions. There have been recent trends for clutchless manual systems, called 'semi-automatic' or 'automated manual', and automatics with an option to change gear manually, sometimes called 'Tiptronic', 'Steptronic' or 'Easytronic'. 'CVT' (continuously variable transmission) refers to an automatic with a single 'speed': the system uses rubber or steel belts to take engine power to the driven wheels, with drive pulleys that expand and contract to vary the gearing. A 'sequential manual' is a manual gearbox with preset gear ratios that are ordered sequentially.

suspension
All suspension systems cushion the car against road or terrain conditions to maximize comfort, safety and roadholding. Heavy and off-road vehicles use 'rigid axles' at the rear or front and rear; these are suspended using robust, leaf-type springs and steel 'wishbones' with 'trailing arms'. 'Semi-rigid axles' are often found at the back on front-wheel-drive cars, in conjunction with a 'torsion-beam' trailing-arm axle. 'Independent' suspension means each wheel can move up and down on its own, often with the help of 'trailing arms' or 'semi-trailing arms'. A 'MacPherson strut', named after its inventor, a Ford engineer called Earl MacPherson, is a suspension upright, fixed to the car's structure above the top of the tyre. It carries the wheel hub at the bottom and incorporates a hydraulic damper. It activates a coil spring and, when fitted at the front, turns with the wheel.

brakes
Almost all modern cars feature disc brakes all round. A few low-powered models still feature drum brakes at the back for cost reasons. 'ABS' (anti-lock braking system) is increasingly fitted to all cars: it regulates brake application to prevent the brakes locking in an emergency or slippery conditions. 'BA' (brake assist) is a system that does this electro-hydraulically, while 'EBD' (electronic brake-force distribution) is a pressure regulator that, in braking, spreads the car's weight more evenly so that the brakes do not lock. 'ESP' (electronic stability programme) is Mercedes-Benz's electronically controlled system that helps keep the car pointing in the right direction at high speeds; sensors detect wayward roadholding and apply the brakes indirectly to correct it. 'Dynamic stability' is a similar system. 'Brake-by-wire' is a totally electronic braking system that sends signals from brake pedal to brakes with no mechanical actuation whatsoever. 'TCS' (traction-control system) is a feature that holds acceleration slip within acceptable levels to prevent wheelspin and therefore improves adhesion to the road. 'VSC' (vehicle stability control) is the computer-controlled application of anti-lock braking to all four wheels individually to prevent dangerous skidding during cornering.

tyres
The size and type of wheels and tyres are given in the internationally accepted formula. Representative examples include: 315/70R17, 235/50VR18, 225/50WR17, 235/40Z18 and

225/40ZR18. In all cases the number before the slash is the tyre width in millimetres. The number after the slash is the height-to-width ratio of the tyre section as a percentage. The letter R denotes radial construction. Letters preceding R are a guide to the tyre's speed rating, denoting the maximum safe operating speed. H tyres can be used at speeds up to 210 km/h (130 mph), V up to 240 km/h (150 mph), W up to 270 km/h (170 mph) and Y up to 300 km/h (190 mph). Finally, the last number is the diameter of the wheel in inches. A 'PAX' is a wheel-and-tyre in one unit, developed by Michelin (for example, 19/245 PAX means a 19 in. wheel with a 245 mm tyre width). The rubber tyre element is clipped to the steel wheel part, rather than held on by pressure. The height of the tyre walls is reduced, which can free up space for better internal packaging, or for bigger wheels for concept car looks. It can also run flat for 200 km at 80 km/h, eliminating the need for a spare.

wheelbase	The exact distance between the centre of the front wheel and centre of the rear wheel.
track front/rear	The exact distance between the centre of the front or rear tyres, measured across the car at the ground.
kerb weight	The amount a car weighs with a tank of fuel, all oils and coolants topped up, and all standard equipment but no occupants.
CO_2 emissions	Carbon-dioxide emissions, which are a direct result of fuel consumption. CO_2 contributes to the atmospheric 'greenhouse effect'. Less than 100 g/km is a very low emission, 150 g/km is good, 300 g/km is bad. 'PZEV' (partial zero emission vehicle) refers to a low-level emission standard that was created to allow flexibility on ZEV standards in California.

Other terms

A-, B-, C-, D-pillars	Vertical roof-support posts that form part of a car's bodywork. The A-pillar sits between windscreen and front door, the B-pillar between front and rear doors, the C-pillar between rear doors and rear window, hatchback or estate rear side windows, and the D-pillar (on an estate) between rear side windows and tailgate. Confusingly, however, some designs refer to the central pillar between front and rear doors as a B-pillar where it faces the front door and a C-pillar where it faces the rear one.
all-wheel drive (AWD)	A system delivering the appropriate amount of engine torque to each wheel via a propshaft and differentials, to ensure that tyre slippage on the road surface is individually controlled. This system is ideal for high-performance road cars, such as Audis, where it is called 'quattro'.
beltline	*See* daylight opening line.
cant rail	The structural beam that runs along the tops of the doors.
coefficient of drag (Cd)	This is shorthand for the complex scientific equation that proves how aerodynamic a car is. The Citroën C-Airdream, for example, has a Cd of 0.28, but the Citroën SM of thirty years ago measured just 0.24, so little has changed in this respect. 'Drag' means the resistance of a body to airflow, and low drag means better penetration, less friction and therefore more efficiency, although sometimes poor dynamic stability.
daylight opening line (DLO)	The line where the door glass meets the door panel, sometimes referred to as beltline or waistline.
diffuser	A custom-designed airflow conduit, often incorporated under the rear floor on high-performance and competition cars, which controls and evenly distributes fast-moving airflow out from beneath the speeding car. This ducting arrangement slows the flow of rushing air behind the car, lowering its pressure and so increasing aerodynamic downforce. The result is improved roadholding.
drive-by-wire technology	Increasingly featured on new cars, these systems do away with mechanical elements and replace them by wires transmitting electronic signals to activate such functions as brakes and steering.
drivetrain	The assembly of 'organs' that gives a car motive power: engine, gearbox, drive shaft, wheels, brakes, suspension and steering. This grouping is also loosely known these days as a 'chassis', and can be transplanted into several different models to save on development costs.

fairing	A sculpted body surface blending different parts of a vehicle together to achieve a streamlined effect.
fast windscreen	A windscreen angled acutely to reduce wind resistance and accentuate a sporty look.
fastback	This refers to the profile of a hatchback that has a rear screen at a shallow angle, so that the tailgate forms a constant surface from the rear of the roof to the very tail end of the car.
feature line	A styling detail usually added to a design to differentiate it from its rivals, and generally not related to such functional areas as door apertures.
flex-fuel	A flex-fuel vehicle (FFV) is an automobile that can accept a variety of different fuels, usually in the same tank. The most common current example is a vehicle that can run on blends of petrol and bioethanol, ranging from 100 per cent petrol to 85 per cent bioethanol–15 per cent petrol. Dual-fuel vehicles carry an additional natural gas tank and can switch between petrol and gas.
four-wheel drive	This refers to a system delivering a car's power to its four wheels. In a typical 'off-road'-type four-wheel-drive vehicle, the differentials can be locked so that all four wheels move in a forward direction even if the tyres are losing grip with the road surface. This makes four-wheel drive useful when travelling across uneven terrain.
glasshouse/greenhouse	The car-design industry's informal term for the glazed area of the passenger compartment that usually sits above the car's waist level.
head-up display	A technology by which useful data are projected upward on to the inside of the windscreen so that information can be displayed in the driver's line of vision.
high-intensity discharge (HID) headlamps	HID headlamps use an electric arc to produce the light, and are also known as xenon headlamps because of the gas used in the lamp. These lamps produce an immediate high-intensity light when switched on.
HVAC	'Heating, ventilation and air-conditioning system.'
hybrid vehicle	A vehicle powered by a combination of two power sources, with a computer program continually deciding the most efficient combination. The most familiar pairing, as on the Toyota Prius, is that of an electric motor for low speeds and a petrol engine for faster driving. An on-board battery is able to store energy recuperated under braking (*see* regenerative braking). Other combinations such as diesel/electric and fuel cell/electric are also possible.
hydrogen fuel cell	A fuel cell produces electricity by combining hydrogen from an on-board tank with oxygen from the atmosphere. The only waste product is water, making fuel-cell vehicles emission-free at the point of use. Practical problems still to be overcome include cold-climate operation, hydrogen storage and the cost of the proton exchange membrane (PEM) at the heart of the cell.
instrument panel	The trim panel that sits in front of the driver and front passenger.
Kamm tail	Sharply cut-off tail that gives the aerodynamic advantages of a much longer, tapering rear end, developed in racing in the 1960s.
monospace/ monovolume/'one-box'	A 'box' is one of the major volumetric components of a car's architecture. In a traditional saloon, there are three boxes: one for the engine, one for the passengers and one for the luggage. A hatchback, missing a boot, is a 'two-box' car, while a large MPV such as the Renault Espace is a 'one-box' design, also known as a 'monospace' or 'monovolume'.
MPV	Short for 'multi-purpose vehicle', this term is applied to tall, spacious cars that can carry at least five passengers, and often as many as nine, or versatile combinations of people and cargo. The 1983 Chrysler Voyager and 1984 Renault Espace were the first. The 1977 Matra Rancho was the very first 'mini-MPV', but the 1991 Mitsubishi Space Runner was the first in the modern idiom.
packaging space	Any three-dimensional zone in a vehicle that is occupied by component parts or used during operation of the vehicle.
platform	Also known as the 'floorpan': the invisible, but elemental and expensive, basic structure of a modern car. It is the task of contemporary car designers to achieve maximum aesthetic diversity from a single platform.

powertrain	The engine, gearbox and transmission 'package' of a car.
regenerative braking	When braking in a hybrid electric vehicle, the electric motor that is used to propel the car reverses its action and turns into a generator, converting kinetic energy into electrical power, which is then stored in the car's batteries.
rotary engine	The rotary engine is very different from a conventional piston engine, being, essentially, a triangular-sectioned shaft that rotates within an elongated chamber to create the compression and combustion cycle. It was developed by Felix Wankel in the 1950s.
shift paddles	A term used for steering-column-mounted levers that, when pulled, send electronic signals to the gearbox requesting a gear change. They were first used in Formula One motor racing.
spaceframe	A structural frame that supports a car's mechanical systems and cosmetic panels.
splitter	Sometimes found at the front of high-performance cars near to ground level, this is a system of under-car ducting that splits the airflow sucked under the car as it moves forward, so the appropriate volume of cooling air is distributed to both radiator and brakes.
sub-compact	You need to rewind fifty-six years for the origins: in 1950, Nash launched its Rambler, a two-door model smaller than other mainstream American sedans. The company coined the term 'compact' for it although, by European standards, it was still a large car. Nash's descendant American Motors then invented the 'sub-compact' class in 1970 with the AMC Gremlin, a model with a conventional bonnet and a sharply truncated hatchback tail; this was quickly followed by the similar Ford Pinto and Chevrolet Vega. In the international car industry today, 'sub-compact' is used as another term for 'A-segment', the range of smallest cars, intended mostly for city driving.
SUV	Short for 'sport utility vehicle', a four-wheel-drive car designed for leisure off-road driving but not necessarily agricultural or industrial use. Therefore a Land Rover Defender is not an SUV, while a Land Rover Freelander is. The line between the two is sometimes difficult to draw, and identifying a pioneer is tricky: SUVs as we know them today were defined by Jeep in 1986 with the Wrangler, Suzuki in 1988 with the Vitara, and Daihatsu in 1989 with the Sportrak. There is also a trend towards more sporty trucks, which has led to the more specific term 'SUT', or 'sport utility truck'.
swage line	A groove or moulding employed on a flat surface to stiffen it against warping or vibration. In cars, swage lines add 'creases' to bodywork surfaces, enabling designers to bring visual, essentially two-dimensional interest to body panels that might otherwise look slab-sided or barrel-like.
Targa	Porsche had been very successful in the Targa Florio road races in Sicily, so, in celebration, in 1965 the company applied the name 'Targa' (the Italian for shield) to a new 911 model that featured a novel detachable roof panel. It is now standard terminology for the system, although a Porsche-registered trademark.
telematics	Any individual communication to a car from an outside base station; this could be, for example, satellite navigation signals, automatic emergency calls, roadside assistance, traffic information and dynamic route guidance.
transaxle	Engineering shorthand for 'transmission axle': this is the clutch and gearbox unit that is connected to the drive shafts to transfer power to the driven wheels. All front-wheel-drive and rear- or mid-engined, rear-wheel-drive cars have some type of transaxle.
tumblehome	The angle of the door glass when viewing a car from the front. The more upright the glass, the less tumblehome.
venturi tunnel	A venturi is an air-management system under a car designed to increase air speed by forcing it through tapered channels. High air speed creates a low-pressure area between the bottom of the car and the road, which in turn creates a suction effect holding the car to the road. Pressure is then equalized in the diffuser at the rear of the car.
waistline	*See* daylight opening line.

Where the New Models Were Launched

New York International Auto Show
14–23 April 2006

Concept
Scion Fuse

Production
Acura MDX
Audi TT
Infiniti G35
Jeep Patriot
Mazda CX-9
Nissan Altima
Saturn Aura
Saturn Outlook
Suzuki XL-7

Busan International Motor Show
28 April – 7 May 2006

Production
Kia Carens

British International Motor Show
28 May – 6 June 2006

Production
Chrysler Sebring
Land Rover Freelander 2
Opel/Vauxhall Corsa

Paris Motor Show
30 September – 15 October 2006

Concept
Citroën C-Métisse
Daihatsu D-Compact X-Over
Ford Iosis X
Lancia Delta HPE
Peugeot 908 RC
Renault Koleos
Renault Nepta
Skoda Joyster
Suzuki Splash
Venturi Eclectic
Volkswagen Iroc

Production
Alfa Romeo 8C Competizione
Audi R8
Citroën C4 Picasso
Dodge Avenger
Kia Cee'd
Mini
Nissan Qashqai
Opel/Vauxhall Antara
Toyota Auris
Volvo C30

Beijing International Auto Show
19–27 November 2006

Concept
Volkswagen Neeza

Production
Nissan Livina Geniss

Greater LA Auto Show
1–10 December 2006

Concept
Acura Advance
Ford Mustang
Honda Remix
Honda Step Bus
Hyundai HCD10 Hellion
Mazda Nagare
Volkswagen Tiguan

Production
BMW X5
Saturn Vue

Motor Show di Bologna
7–17 December 2006

Production
Daihatsu Materia

North American International Auto Show (NAIAS)
13–21 January 2007

Concept
Acura Advanced Sports Car
Chevrolet Volt
Chrysler Nassau
Ford Airstream
Ford Interceptor
Honda Accord Coupé
Jaguar C-XF
Jeep Trailhawk
Kia Kue
Lincoln MKR
Mazda Ryuga
Mercedes-Benz Ocean Drive
Nissan Bevel
Toyota FT-HS
Volvo XC-60

Production
Cadillac CTS
Chevrolet Malibu
Chrysler Town and Country
Ford Focus
Hyundai Veracruz
Mazda Tribute
Mitsubishi Lancer
Nissan Rogue
Toyota Tundra

Chicago Auto Show
9–18 February 2007

Production
Pontiac G8
Scion xB
Scion xD
Toyota Highlander

Geneva International Motor Show
8–18 March 2007

Concept
Bertone Barchetta
Dodge Demon
EDAG LUV
Fioravanti Thalia
Honda Small Hybrid Sports
Hyundai Quarmaq
Italdesign Vadho
KTM X-Bow
Lada C-Class
Mazda Hakaze
Opel GTC
Rinspeed eXasis
Stola Coupé
Tata Elegante
Toyota Hybrid X
Zagato Diatto Otto Vu

Production
Audi A5
Bentley Brooklands
Citroën C-Crosser
Fiat Bravo
Ford Mondeo
Hyundai i30
Lotus 2-Eleven
Maserati GranTurismo
Mazda2
Mercedes-Benz C-Class
Nissan X-Trail
Peugoet 4007
Renault Twingo
Skoda Fabia
Smart Fortwo
Tramontana
Volvo V70

Major International Motor Shows 2007–2008

Middle East International Motor Show
14–18 November 2007
Dubai World Trade Center, Dubai, UAE
dubaimotorshow.com

Greater LA Auto Show
16–25 November 2007
Los Angeles Convention Center, Los Angeles, USA
laautoshow.com

Motor Show di Bologna (Salone Internazionale dell'Automobile)
7–16 December 2007
BolognaFiere, Bologna, Italy
motorshow.it

Riyadh Motor Show
9–13 December 2007
Riyadh Exhibition Center, Riyadh, Saudi Arabia
recexpo.com

Chicago Auto Show
Date not yet available (February 2008)
McCormick Place South, Chicago, USA
chicagoautoshow.com

Canadian International Auto Show
15–24 February 2008
Metro Toronto Convention Center and SkyDome, Toronto, Canada
autoshow.ca

Melbourne International Motor Show
28 February – 10 March 2008
Melbourne Exhibition Center, Melbourne, Australia
motorshow.com.au

Geneva International Motor Show
6–16 March 2008
Palexpo, Geneva, Switzerland
salon-auto.ch

New York International Auto Show
21–30 March 2008
Jacob Javits Convention Center, New York, USA
autoshowny.com

Budapest Auto Show
Date not yet available (May 2008)
HUNGEXPO Budapest Fair Centre, Budapest, Hungary
automobil.hungexpo.hu

British International Motor Show
23 July – 3 August 2008
Excel, London, UK
britishmotorshow.co.uk

Moscow International Motor Show
Late August – early September 2008
Crocus Expo, Moscow, Russia
motorshow-ite.com

Frankfurt International Motor Show
Date not yet available (September 2008)
Trade Fairgrounds, Frankfurt am Main, Germany
iaa.de

Paris Motor Show
4–19 October 2008
Paris Expo, Paris, France
mondial-automobile.com

Prague Auto Show
18–21 October 2007
Prague Exhibition Grounds, Prague, Czech Republic
incheba.cz

Tokyo Motor Show
27 October – 11 November 2007
Nippon Centre, Makuhari, Chiba, Tokyo, Japan
tokyo-motorshow.com

Marques and Their Parent Companies

Hundreds of separate carmaking companies have consolidated over the past decade or so into the following groups: BMW, DaimlerChrysler, Fiat, Ford, General Motors, Honda, Hyundai, Proton, PSA–Peugeot Citroën, Renault–Nissan, Toyota and Volkswagen. These account for at least nine out of every ten cars produced globally today. The remaining independent marques either produce specialist models, offer niche design and engineering services or tend to be at risk because of their lack of economies of scale. The global over-capacity in the industry means that manufacturers are having to offer increased choice to the consumer to differentiate their brands and maintain market share. Not all parent companies fully own the carmakers listed as under their control. Mazda, for example, is a key part of the Ford Alliance but is only one-third owned by the US giant.

Note: As we close for press we have received news that DaimlerChrysler has sold its US Chrysler division to private-equity buyer Cerberus. The German business will be known as Daimler AG, and the North American operations will become Chrysler Holding LLC. Details of the effects on the group's various brands and models will become clear once the split becomes effective later in 2007.

BMW
BMW
Mini
Riley*
Rolls-Royce
Rover*
Triumph*

DaimlerChrysler
Chrysler
De Soto*
Dodge
Hudson*
Imperial*
Jeep
Maybach
Mercedes-Benz
Nash*
Plymouth*
Smart

Fiat Auto
Abarth*
Alfa Romeo
Autobianchi*
Ferrari
Fiat
Innocenti*
Lancia
Maserati

Ford
Aston Martin
Daimler*
Ford
Jaguar
Lagonda*
Land Rover/ Range Rover
Lincoln
Mazda
Mercury
Volvo

General Motors
Buick
Cadillac
Chevrolet
Corvette
Daewoo
GM
GMC
Holden
Hummer
Oldsmobile*
Opel
Pontiac
Saab
Saturn
Suzuki
Vauxhall

Honda
Acura
Honda

Hyundai
Asia Motors
Hyundai
Kia

Proton
Lotus
Proton

PSA–Peugeot Citroën
Citroën
Hillman*
Humber*
Panhard*
Peugeot
Simca*
Singer*
Sunbeam*
Talbot*

Renault–Nissan Alliance
Alpine*
Dacia
Datsun*
Infiniti
Nissan
Renault
Renault Sport
Samsung

Toyota
Daihatsu
Lexus
Scion
Toyota
Will*

Volkswagen Group
Audi
Auto Union*
Bentley
Bugatti
Cosworth
DKW*
Horch*
Lamborghini
NSU*
SEAT
Skoda
Volkswagen
Wanderer*

Independent marques
Austin-Healey*
AviChina
Beijing
Bertone
Bristol
Byd
Caterham
Chery
Dongfeng
Donkervoort
EDAG
Elfin
ETUD
Farboud
Fenomenon
Fioravanti
Fisker Coachbuild
Heuliez
Hindustan
Inovo
Invicta
Irmscher
Isuzu
Italdesign
Izh
Jensen
Joss
Koenigsegg
KTM
Lada
Mahindra
Marcos
Maruti
Mitsubishi
Mitsuoka
Morgan
Nanjing (MG*)
Pagani
Panoz
Paykan
Perodua
Pininfarina
Porsche
Rinspeed
SAIC/Roewe (Austin*, Morris*, Wolseley*)
Sivax
Spyker
SsangYong
Stola
Subaru
Tata
Th!nk
Tramontana
TVR
Venturi
Volga
Westfield
Wiesmann
Zagato
ZAZ
ZIL

* Dormant marques

Acknowledgements

This book could not have been written without the help and support of a number of people. I would like to extend a special thank you to the team at Merrell Publishers for their professional work in helping to create another edition in the *Car Design Yearbook* series. Particular thanks go to Marion Moisy, Kirsty Seymour-Ure, Paul Shinn, John Grain and Michelle Draycott.

Thanks are also due to Tony Lewin for bringing his extensive and specialist knowledge to the editorial team, and to the manufacturers themselves for supplying information and photographs. Finally, a huge thank you to Hannah for living through every model.

Stephen Newbury
Henley-on-Thames, Oxfordshire
2007

Picture Credits

The illustrations in this book have been reproduced with the kind permission of the following:

Acura
Alfa Romeo
Audi AG
Bentley
Bertone
BMW AG
Cadillac
Chevrolet
Citroën
Daihatsu
DaimlerChrysler
Dodge
EDAG
Fiat Auto
Fioravanti
Ford Motor Company
Honda Motor Co.
Hyundai Car UK
Infinity
ItalDesign
Jaguar Cars
Jeep
Kia Motors Corporation
KTM
Lada
Lancia
Land Rover
Lincoln
Lotus Cars
Maserati
Mazda Motors
Mercedes-Benz
Mini
Mitsubishi Motors Corporation
Nissan Motors
Opel AG
Peugeot SA
Pontiac
Renault SA
Rinspeed
Saturn
Scion
Skoda
Smart
Stola
Suzuki Motor Corporation
Tata Motors
Toyota Motor Corporation
Tramontana
Venturi
Volkswagen AG
Volvo Car Corporation
Zagato

Giles Chapman Library

Pages 20–25: Pilkington
Page 26: Photo Nuon/
Hans-Peter van Velthoven

MERRELL

First published 2007 by Merrell Publishers Limited

Head office
81 Southwark Street
London SE1 0HX

New York office
740 Broadway, Suite 1202
New York, NY 10003

merrellpublishers.com

British Library Cataloguing-in-Publication data:
Newbury, Stephen
The car design yearbook 6 : the definitive annual guide to all new concept and production cars worldwide
1. Automobiles – Periodicals 2. Automobiles – Design – Periodicals
I. Title
629.2′22′05

ISBN-13: 978-1-8589-4378-7
ISBN-10: 1-8589-4378-7

Consultant editor: Tony Lewin
Copy-edited by Kirsty Seymour-Ure
Proof-read by Barbara Roby
Designed by John Grain
Design concept by Kate Ward

Printed and bound in Singapore

Frontispiece: Mazda Ryuga
Pages 6–7: Peugeot 908 RC
Pages 18–19: Ford Airstream
Pages 34–35: Audi TT
Pages 250–51: Land Rover Freelander 2
Pages 270–71: Jaguar C-XF